Constraints and Advanced Solutions for LTE-Advanced Heterogeneous Networks

This book presents advanced techniques for enhancing mobility management across evolving wireless systems, including 5G and beyond. It introduces innovative handover decision algorithms that leverage user movement patterns, historical data, and multi-criteria optimization to improve the reliability and efficiency of both horizontal and vertical handovers. Focusing on dense femtocell–macrocell environments, the book proposes refined handover parameters and distance-based vertical handover methods for small-cell transitions. It addresses mobility challenges faced by high-mobility users, such as in vehicular and intelligent transportation systems, by introducing trajectory-based handover prediction models. The practical applications of these techniques are explored within the contexts of smart cities, Internet of Things (IoT) ecosystems, and future network infrastructures.

Features

- Introduces novel techniques to optimize horizontal and vertical handovers, addressing key challenges in dense and heterogeneous networks.
- Proposes refined parameters for reliable handover performance in femtocell, macrocell, and small-cell environments.
- Enhances target cell selection strategies using trajectory-based predictions for users in rapid motion.
- Develops distance-based vertical handover methods to improve handovers for low-mobility users within small-cell deployments.
- Explores real-world use cases across intelligent transport systems, urban infrastructure, and IoT networks.
- Aligns proposed solutions with next-generation wireless standards and mobility requirements.

This book is a practical guide and essential reference for researchers, engineers, and students in wireless communications, networking, and mobility management.

Constraints and Advanced Solutions for LTE-Advanced Heterogeneous Networks

Rami Ahmad, Elankovan A. Sundararajan and
Mohammad Kamrul Hasan

CRC Press
Taylor & Francis Group
Boca Raton London New York

CRC Press is an imprint of the
Taylor & Francis Group, an **informa** business
A CHAPMAN & HALL BOOK

Designed cover image: ShutterStock

First edition published 2026
by CRC Press
2385 NW Executive Center Drive, Suite 320, Boca Raton, FL 33431

and by CRC Press
4 Park Square, Milton Park, Abingdon, Oxon, OX14 4RN

CRC Press is an imprint of Taylor & Francis Group, LLC

ISBN: 9781041086697 (hbk)
ISBN: 9781041090380 (pbk)
ISBN: 9781003648130 (ebk)

DOI: 10.1201/9781003648130

Typeset in Times
by Newgen Publishing UK

Contents

Preface

The Heterogeneous Network (HetNet) is considered a promising future solution. This network uses new cells deployment system, which is known as the two-tier LTE-A network. The LTE-A wireless network system depends purely on the hard handover (HHO) procedure. HHO is standardized via 3GPP (Release 8 to 13) to allow user equipment (UE) to migrate among the cells. To trigger the procedure, horizontal and vertical handover decisions typically depend on Received Signal Strength and hysteresis parameters. HHO decision reduces wireless network complexity and enhances high-velocity user reliability. However, due to the effect of interference, this scenario could still result in high handover probability and throughput degradation.

Interference arises from the high-density femtocells in Received Signal Quality. Moreover, other issues also affect handover performance. For example, a long time is needed to select the appropriate target cell. Besides that, due to high mobility, the switching between horizontal/vertical handovers becomes inefficient. The main objective of this research is to propose and implement an efficient handover decision procedure based on user profile in an LTE-A HetNet. Firstly, new multi-criterion handover decision parameters are proposed in typical/dense femtocells in macrocells and the cell–edge environment. Secondly, the UE Moving Direction and its Informative Historical Handover Decision (MDIH-HD) algorithm is introduced for horizontal networks. This is based on the UE trajectory and the high-mobility UE routine route to estimate the target cell in a short time. Thirdly, the low-mobility UE Moving Direction and Distance Vertical Handover Decision (MDD-VHD) algorithm between the UE and small cells is proposed to estimate the target small cell for handover. Finally, new handover decision parameters are investigated to select their threshold values. The proposed handover algorithms are validated using the LTE-Sim simulators under an urban environment. The simulation results are encouraging. A reduction in the average number of handovers, the average number of signaling measurements, and packet loss and delay was observed. At the same time, the system throughput was increased. The MDIH-HD reduced the handover number, packet delay, and packet loss by 18%, 12%, and 47%, respectively, whereas the system throughput increased by 0.54% and 0.33% compared to conventional and reference algorithms, respectively. Meanwhile, the MDD-VHD reduced the number of handovers, packet delay, and packet loss by 48%, 87%, and 82%, respectively. At the same time, the system throughput in the two-tier network increased by 14.3% compared to reference work.

There are seven chapters in this book including the current Chapter 1. This chapter explores the research background in general, the research problems, objectives, and challenges. The rest of the book is organized as follows:

Chapter 2 provides a background of the LTE-A system and the evolution of LTE and LTE-A. It also provides a detailed overview of the LTE-A structure. Then, an overview of the handover is provided, including both horizontal and vertical decision procedures, handover parameters, and handover failures. Next, a brief description of handover procedures, which include horizontal and vertical procedures, is given. Lastly, related research works are presented.

Chapter 3 explains the generic approach for enhancements used for the proposed solutions in this subject domain of this book. A general description of the approach for enhancement is given, followed by a brief description of the proposed solutions. Then, the simulation model and simulation flowchart used to validate the proposed solutions are described. Then, the system performance metrics used to characterize network service quality are defined. Finally, the input parameters setting is discussed for both proposed algorithms to get the threshold values.

Chapter 4 focuses on the efficient handover decision algorithm in the horizontal networks to further enhance system performance when UE mobility velocity is high. Firstly, this chapter investigates and discusses the effectiveness of the proposed parameters in MDIH-HD. Then, the algorithm performance is presented and discussed in terms of average handover number, average signaling measurement number, packet delay ratio, packet loss ratio, and system throughput. Then, the MDIH-HD algorithm is also compared to existing works.

Chapter 5 focuses on the efficient handover decision algorithm in the vertical network to further enhance system performance when UE mobility velocity is low. Firstly, this chapter investigates and discusses the effectiveness of the proposed parameters. Then, the MDD-VHD performance is presented and discussed in terms of average number of handovers, average number of signaling measurements, packet delay ratio, packet loss ratio, and system throughput. Then, the MDD-VHD algorithm is also compared to existing works.

Chapter 6 introduces the application of the proposed handover algorithms in smart city environments. This chapter explores how LTE-A heterogeneous networks support smart city applications, including Intelligent Transportation Systems, Smart Healthcare Systems, Autonomous Vehicle Communication, Smart Energy Management, and Public Safety and Emergency Services. It also discusses optimizing vertical handovers for IoT applications and addressing network congestion and interference challenges in urban areas.

Finally, Chapter 7 concludes this book, and also presents, in brief, the research contributions and a few significant recommendations for future work.

Abbreviations

1G	First generation
2G	Second generation
3G	Third generation
3GPP	Third Generation Partnership Program
4G	Fourth generation
5G	Fifth generation
AMPS	Advanced Mobile Phone System
ANDSF	Access Network Discovery and Selection Function
ANH	Average number of handovers
ANR	Automatic neighbor relation
ASMN	Average of Signaling Measurement Number
BCCH	Broadcast Control Channel
CA	Carrier aggregation
CBR	Constant bit rate
CDMA	Code Division Multiple Access
CGI	Cell Global Identifier
CIR	Carrier Interference Ratio
CN	Core Network
CoMP	Coordinated MultiPoint
CP	Cyclic prefix
CSG	Closed Subscriber Group
DL	Down Link
DSSS	Direct Sequence Spread Spectrum
E-UTRAN	Evolved UMTS Terrestrial Radio Access Network
eNB	evolved NodeB
EPC	Evolved Packet Core
FDD	Frequency Division Duplex
FDMA	Frequency Division Multiple Accesses
GSM	Global System for Mobile communications
HeNB	Home-eNB
HeNB-GW	HeNB gateway
HetNet	Heterogeneous network
HF	Handover failure
HHO	Hard handover
HO-Request	Handover-request
Hys	Hysteresis
IMT-Advanced	International Mobile Telecommunications Advanced
IPs	Internet Protocols
IS-95	Interim Standard system
ITU	International Telecommunication Union
ITU-R	International Telecommunication Union Radio communication

LTE	Long-term evolution
LTE-A	Long-term evolution advanced
MDD-VHD	Moving Direction Distance Vertical Handover Decision
MDIH-HD	Moving Direction Informative Historical Handover Decision
MIMO	Multiple inputs multiple outputs
MME	Mobility management entity
NAS	Non-access-stratum
NLOS	Non-line-of-sight
OFDMA	Orthogonal frequency-division multiple access
PCI	Physical cell identifier
PCRF	Policy and charging rules function
PDN-GW	Packet Data Network Gateway
PDR	Packet delay ratio
PLMN	Public Land Mobile Network
PLR	Packet loss ratio
QoS	Quality of Service
RAT	Radio access technology
RB	Resource block
RLF	Radio link failure
RN	Relay node
RNC	Radio network controller
RRM	Radio resource management
RSQ	Received signal quality
RSRP	Reference signal received power
RSS	Received signal strength
RSSI	Received signal strength indicator
RTP	Real-time Transport Protocol
S-GW	Serving gateway
S1AP	S1 Application Part
SAE	System architecture evolution
SC-FDMA	Single-carrier frequency division multiple access
SI	System information
SINR	Signal interference plus noise ratio
SN	Sequence number
SNR	Signal-to-noise ratio
TACS	Total Access Communication System
TAU	Tracking area update
TD-SCMA	Time division synchronous code division multiple access
TDMA	Time division multiple access
TH	Throughput
TTT	Time to trigger
UDP	User Datagram Protocol
UE	User equipment
UL	UpLink
UMTS	Universal Mobile Telecommunications System

VNI	Visual network index
VoIP	Voice over Internet Protocol
WCDMA	Wide code division multiplexing access
WiMAX	Worldwide interoperability for microwave access
WLAN	Wireless Local Area Network
WWAN	Wireless Wide Area Network
X2-GW	X2-Gateway
X2AP	X2 Application Part

Symbols

$2E^2$	The mean power of multipath received signal power
α	The maximum angle of UE direction
β	The angle of UE direction (slope)
β_{Thd}	The slope threshold
μ_{dB}	The mean value of received signal power in dB
ω	The transmitted-to-received power ratio
σ_{dB}	Standard deviations in dB
θ	The angle of UE direction
$^{\theta}c$	The angle between UE current position, UE previous position, and cell location for each cell
^{A}Pr	The amplitude of the received signal power
$Anorm$	The angle normalization
BS_T	Target base station
BSs	Serving base station
C	The position of each neighbor tower cell
c	The index of the cell
D	Antenna height in meters
dc	The distance between UE current location and each cell tower position
$d_{threshold}$	Distance threshold
$delay_{ki}$	The total packet delay of UE_i at its serving $cell_k$
E	The total packets delay of UEi at its serving $cell_k$
Dnorm	The distance normalization
eNB_s	Serving eNB
G	The channel gain between the user and the cell
Hys_s	The hysteresis parameter of serving cell
j	The last UE position
L	The path loss between UE and cell
L_{fad}	Shadow fading with a log-normal distribution and a standard deviation
N	The total number of cells
n	The index of the PRB
^{N}HO	The total number of handovers
^{N}SM	The total number of signaling measurements
O	An empty array
P	The instantaneous signal power received by UE for each node
P_1	The previous UE position
P_2	The current UE position
P_i	The sum of the powers received by the UE, coming from the neighbor cells
$^{P}noise$	The variance of the additive white Gaussian noise
$^{P}T\,hd$	The UE current position threshold

$Pack_r$	The total number of received packets
$Pack$	The total number of packets that is created through simulation time
Ps	Position between previous and newest UE positions
Q	The Queue is used to trace the newest UE positions
$Q(last)$	The UE current position
Q_n	The RSRP of the neighbor cell
Q_s	The RSRP of serving cell
R	Radius of cell in km
r	Radius of femtocell in meter
$^{RSS}threshold$	Received signal strength threshold
S	Distance threshold value between P_1 and P_2
^{S}Thd	The distance threshold
T	Simulation time
t	Timeslot
$^{T}measurement$	Measurement interval time
Tb	The UE table
TH_{ki}	The total size of packet correctly received by UE_i
u	Index of the UE
V	The UE velocity
W	The summation result value

Author Biography

Rami Ahmad is currently working as Associate Professor of Computer Science, Networking, and Network Security at the American University in the Emirates. Dr Ahmad is an academic and researcher in the field of wireless networks and sensor networks—focusing primarily on management, security and lightweight security techniques, as well as network self-management and various applications of the network embedding system. Furthermore, he has experience in writing and evaluating applied research proposals in European funding organizations. Dr Ahmad obtained his PhD in Computer Science from the Networking and Security Lab of the National University of Malaysia. He also has a Master's degree in Computer Science in the field of Intelligent Systems from Utara Malaysia University. Dr Ahmad's current research interests include machine learning, computer and wireless networking, network security, lightweight security technologies, and network sensing.

Elankovan A. Sundararajan is Associate Professor at the Faculty of Information Science and Technology, National University of Malaysia (UKM). He received his BSc (Hons) in 1995, MSc in Computer Science in 1997 from the National University of Malaysia, and PhD in Computer Science from the University of Melbourne, Australia in 2008. His current research interests include high-performance computing, big data, cloud computing, and computer networks. He has been a member of the Institute of Electrical and Electronics Engineers since 2005.

Mohammad Kamrul Hasan is currently working as Associate Professor in the Network and Communication Technology research cluster, Center for Cyber Security, Universiti Kebangsaan Malaysia (UKM). He completed his Doctor of Philosophy (PhD) degree in Electrical and Communication Engineering from the Faculty of Engineering, International Islamic University Malaysia, in 2016. Dr Hasan completed his MSc degree in Communication Engineering in the area of Electrical and Communication Engineering from International Islamic University Malaysia in 2012. He has been experienced working in electrical and communication industries, after completing his BSc degree back in 2003. His undertakings included supervision, installation, and configuration of the corporate information-centric network, servers, Wide Area Network (WAN), Wireless Local Area Network (WLAN), Local Area Network (LAN), fault handling, and programming of the ladder logic design for the industrial LAN-connected power line communication (PLC) devices. He has demonstrated excellent research outcomes and received several awards including gold medal, scholarships, and best paper award. He is well versed, specialized with elements pertaining to cutting-edge information-centric network, wireless communication and networking, microcontroller-based electronic devices, smart grid WAM systems, microprocessor systems, and IoT. Dr Kamrul is Senior Member of the Institute of Electrical and Electronics Engineers (IEEE) since 2013, Member of the Institution of Engineering and Technology, and a member of Internet Society. He has served the IEEE IIUM student branch as chair from 2014 to 2016. He has actively participated in many events/workshops/training for the IEEE and IEEE humanity programs in Malaysia. He reviews SCIE journals including *IEEE Access*, *Telecommunication Systems*, Hindawi, SAGE, *ETRI Journal*, and *KSII Transactions on Internet and Information Systems*.

1 Mobility Management in LTE-Advanced Wireless System

Advances and Constraints

1.1 BACKGROUND

In the last two decades, an obvious increase in the demand for wireless communications was observed. This saw an exponential rise in the use of smart equipment devices. The Cisco Visual Network Index global mobile data traffic forecast showed a tremendous rate of increase in broadband traffic up until the year 2020 (Cisco 2017). Cellular network systems are evolving to meet future prerequisites of the coverage area, capacity, and data rate. One such solution is the interoperability between homogeneous and heterogeneous networks. The Third Generation Partnership Program (3GPP) released a long-term evolution (LTE) and LTE-Advanced (LTE-A) network, new radio access wireless networks, to meet these objectives (Kanchi et al. 2013). The new radio access technology supports high-velocity data communication in the homogeneous network. It also supports small cells (femtocells) in the Heterogeneous Network (HetNet). Therefore, better indoor/outdoor coverage in the LTE-A can be achieved. Hence, because LTE-A technology must support high-velocity mobility, mobility improvement is an essential feature of the network (Márquez-Barja et al. 2011).

In wireless networks, each cell or base station has a limited coverage space. To cover all areas, the base stations intersect with each other in the overlapped area. Furthermore, in order to keep the User Equipment (UE) connection active through roaming, the handover procedure is used. The handover procedure, also known as a handoff, is an important strategy. This procedure permits the UE connected with the serving-based station to exchange with the neighbor-based station (target) without session separation (Miyim et al. 2014). Subsequently, mobility, security, connection reliability, and high Quality of Service (QoS) are some advantages of a seamless handover. LTE-A features support UE mobility in high velocities. The network also supports two-tier deployment cells (macro and femto), which are used to deploy coverage area in a space. The macro cell covers a wide area. These macro cells are referred to as evolved NodeB (eNB) in the LTE-A. eNB is used for outdoor coverage. The femtocell is called Home-eNB (HeNB). HeNB covers a small area

DOI: 10.1201/9781003648130-1

and is used to increase the data rates in indoor coverage (Gódor et al. 2015; Salman et al. 2014).

The aforementioned LTE-A features will increase the efficiency of the wireless network system. However, the handover process between the (eNB and HeNB) cells remains a dilemma. Furthermore, the LTE-A wireless network uses the Hard Handover (HHO) procedure for devices roaming between the eNB and HeNB networks. This is because the soft handover has high complexity and utilized in the third-generation (3G) wireless networks. Utilizing HHO implies that the interruption in handover time, buffering, data loss, and delay remain in the same UE plane. In addition, data forwarding would be within distributed controllers in the HHO instead of being in centralized controllers as per the soft handover (Chang et al. 2009; Sipila et al. 1999). In this manner, the handover procedure would be more redundant and rapid. Besides that, the handover performance will turn out mainly in real-time applications. In line with this, the seamless handover scheme should be redundant and fast, and data exchange should not be lost or delayed (Márquez-Barja et al. 2011; Shayea et al. 2012).

There are two types of handover procedures in the LTE-A. The first is the vertical (heterogeneous) handover, which is utilized in the handover between various types of wireless network cells (Ahmad, Sundararajan & Khalifeh 2020). The second is the horizontal (homogeneous) handover, which is utilized in the handover between similar types of wireless network cells (Márquez- Barja et al. 2011). HHO disadvantages include carrier interference, high data loss, and difficulty in keeping up with QoS necessities due to the delay in handover during network cell migration (Chang et al. 2009). Despite this setback, the HHO is useful for UEs because it minimizes the complexity of wireless networks and is reliable for high-velocity users. However, the HHO procedure also results in packet drops. This in turn may reduce the outcome of lost data during an active session. Thus, new techniques must be applied to avoid data loss.

The HHO procedure is carried out in three primary stages: collection of information, decision, and execution (Zhou & Ai 2014). In the first stage, information associated with wireless network cells and UEs is gathered for the handover triggering process. Subsequently, this information is utilized to assess the candidate wireless cells in the decision stage. After selecting the target network cell in the previous stage, a new connection is built up in the execution stage. Then, all the resources are forwarded from the old network cell (serving cell) to the new network cell (target cell) in the Evolved UMTS Terrestrial Radio Access Network (E-UTRAN). Moreover, all Internet Protocols (IPs) and routings will be categorized under mobility protocols in the Evolved Packet Core (EPC). The EPC supports UE connectivity to a multitude of cells (Volkan et al. 2014). However, the long searching time needed for nominating and selecting the target cell for handover minimizes the efficiency of this type of handover.

The HHO procedure in LTE-A uses three main parameters, which control the performance of the handover process and affect the handover decision at the triggering point. The first parameter is the Received Signal Strength (RSS), which refers to the power of signals transmitted by a particular cell. These signal values use a synonymous path loss between the UE and several candidate networks cells (3GPP 2011b). The RSS parameter belongs to the Down Link (DL) measurement channel, which is scanned periodically in the UE site. Next, this parameter is sent to its serving network cell as a measurement report message when the handover procedure occurs (Isa et al. 2015). The second parameter is Hysteresis (Hys), which is also known as the handover margin. The Hys parameter is used to control the handover procedure. This parameter makes the RSS value of the UE of the serving cell stronger than the RSS values of the neighbor cells in real signal values. Thus, the RSS values of the neighbor cells will be sufficiently strong for the UE to start the handover decision (Isa et al. 2015). The last parameter is Time to Trigger (TTT). The threshold time interval allows handover triggering based on this threshold value (Kim et al. 2014).

The general architecture of the LTE-A system together with the handover technique is described in Figure 1.1. These parameter values affect the handover decision stage and the handover performance either in a decreasing or in an increasing manner.

The standard handovers (horizontal and vertical) in LTE-A are based on three parameters. Network operators manually set the values of these parameters. The main drawback of the standard handover is its dependency on RSS measurements in the handover decision regarding all types of network cells. For example, in urban areas, the RSS of eNB deployment is fairly good. The UE will not measure the existing small cells because the signal strength of the serving eNB is still desirable (Chang et al. 2012). Therefore, the RSS parameter is not enough for handover procedure in HetNets. Moreover, the constant threshold of the other parameter values also reduces the effectiveness of the handover procedure. The performance of the handover decision is based on the right time and the precise measurement of handover decision parameters. This is to maintain data connection when the UE is transferred from the serving cell to target cell. Thus, both the density of HeNBs in a particular area and the UE velocity poses new challenges. These include the number of measurement reports and the number of unnecessary handovers (Gódor et al. 2015; Sung et al. 2013; Xenakis et al. 2014). Therefore, in LTE-A wireless networks, the handover decision in the standard procedure must be updated using several input parameters along with the RSS. This way, the performance of the handover decision can be improved.

As previously noted, the handover procedure performance is affected by unnecessary handoffs and UE velocity. Thus, this research proposes a new handover procedure with additional parameters to enhance handover performance using LTE-A technology. This is done through solving the aforementioned shortage in the information-gathering stage and optimizing the handover decision process in both the horizontal and vertical networks. The UE moving direction, UE velocity, and the distance between the UE and candidate cells are considered together with the RSS as input parameters. These input parameters will assist the handover decision in selecting the target cell in a shorter time. Moreover, discovering the UE direction will also assist the handover decision through avoiding measurement controls of the neighbor cells that are not in front of the UE trajectory. The UE velocity aids the handover decision by preventing small cells from becoming candidate cells when the UE moves at high

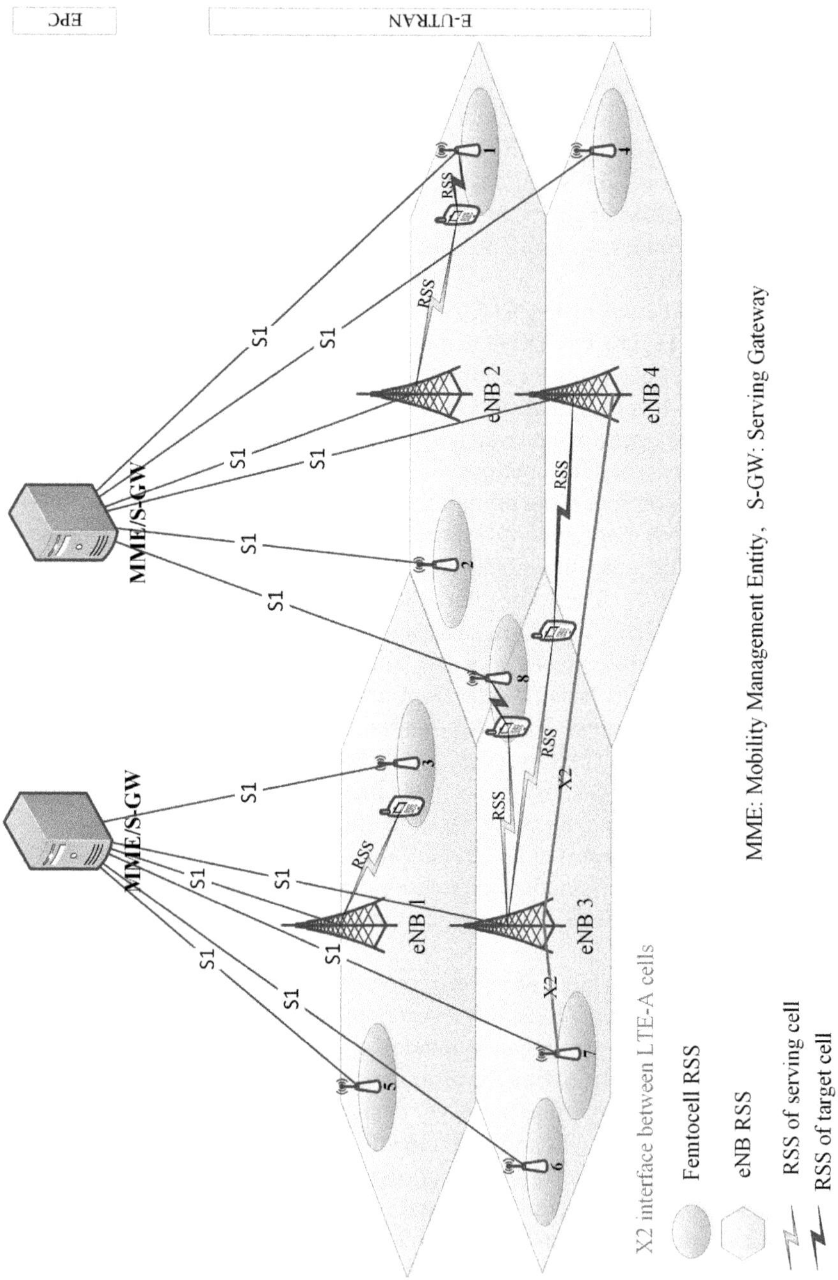

FIGURE 1.1 LTE-A system architecture with handover technique.

velocities. Furthermore, the distance aids the handover decision through selecting the target cell from the candidate cells depending on the proximity between cell tower and the UE current location.

1.2 CONSTRAINTS

Dependency on using the HHO procedure in the LTE-A for UE mobility results in the reduction in wireless network complexity. Additionally, the reliability of high-velocity handovers is increased. In spite of these advantages, the HHO still experiences malfunctions. For example, the HHO decision is affected by carrier interferences and inflexibility. This is to keep up with the QoS requirements in high packet delay and loss during handoff among cells (Chang et al. 2009).

LTE-A supports UE velocities of up to 500 km/h. Therefore, the efficiency of the wireless system will be reduced. Also, the handover process will become more critical when the UE starts moving from the serving cell to the target cell (Chen et al. 2015a; Fei & Fan 2012). Furthermore, the seamless handover procedure should be fast and frequent, and the data transfer should not be delayed and lost. Even more, the increasing demands for using wireless broadband have led to the mobile network operators deploying more wireless cells of various types (eNB and HeNB) to fulfill the data traffic rate. Thus, a handover between the two-tier networks will cause an unnecessary handover effect, handover failure, and a drop in the performance of the wireless network.

The standard handover decision in LTE-A is based on the RSS and Received Signal Quality (RSQ) parameters. Thus, the deployment of dense HeNBs inside the eNB cell would increase the interference. This interference reduces the RSQ by decreasing the RSS ratio in comparison to the Received Signal Strength Indicator (RSSI). As a result, a degradation in handover performance is observed (Lopez-Perez et al. 2011). Moreover, in the case where UE mobility is high, the HHO decision, which is based only on the RSS, will require a longer searching time to determine the target cell. As a result, the handover performance will undergo increased degradation (Márquez-Barja et al. 2011; Shayea et al. 2012). In addition, the variation in transmitting power between the two types of wireless networks (the eNB and HeNB) is another issue. When the HeNB location is near the eNB tower, the RSS of the HeNB will always be less than the eNB. Thus, the main objective of this study, which is to reduce time delay and packet loss in the handover between the network cells when the UE moves at different velocities, will be achieved.

Many research studies have worked on enhancing the handover procedure in both the vertical and horizontal wireless networks in LTE/ LTE-A. This has been achieved by reducing packet loss and packet delay and simultaneously increasing network throughput. LTE/LTE-A promises improvement in handover performance and guarantee of a satisfactory QoS for real-time applications. However, there are still many concerns regarding the capability of adopting these schemes in real environments. This real environment deployment issue can be solved by minimizing the extra overhead of corresponding mechanisms and thus can minimize calculation time.

In summary, the problems that must be addressed in an LTE-A system are listed as follows:

1. Current selection of handover decision parameters is based on RSS and RSQ parameters, which are inappropriate for high-density HeNB deployment in LTE-A networks.
2. Long searching time/frequent handovers due to the presence of multiple target cells in the case of high UE velocity.
    a. The HHO decision based only RSS will take extra searching time to determine the target cell and that also causes other degradation of handover performance.
3. Inefficient switching between horizontal/vertical networks due to high mobility:
    a. Unnecessary handovers effect.
    b. Handover failure effect.
    c. Drop the network performance.

1.3 PROSPECTS OF ENHANCEMENTS

The prospects of enhancements to overcome the constraints are possible improvements in handover technique for horizontal wireless networks (eNB to eNB) in the case of high UE velocity and handover technique for vertical networks (eNB to HeNB) in the case of low UE velocity. Therefore, this book focuses on the prospects of enhancement with the following objectives:

1. To propose a generalized multi-criterion handover decision parameter setting in typical/dense HeNBs in the eNB-cell and users distributed in the cell environment.
2. To implement target cell selection mechanism depending on the UE profile. This mechanism reduces the searching time when selecting the appropriate cell.
3. To implement an efficient handover procedure for switching between the horizontal and vertical LTE-A wireless networks (eNB and HeNB).

2 Handover Management in LTE-A Heterogeneous Networks

2.1 INTRODUCTION

This chapter presents a brief background of the LTE-Advance wireless system. It also focuses on a major issue in the network field, which involves the handover in both horizontal and vertical networks. The fundamental objective of this chapter is to conduct a comprehensive survey regarding research development on the handover procedure. These results are used to identify crucial problems in various handover decisions, as proposed in the literature. A summary description of wireless network evolution is presented in Section 2.2. In Section 2.3, an overview of the LTE-A architecture is briefly clarified. Additionally, the LTE-A core network control plane and LTE-A access network (E-UTRAN) are described. Section 2.4 focuses on the heterogeneous networks in LTE-A and the use of femtocells. Following that, in Section 2.5, a general overview of the handover is presented. This includes definitions of the techniques, classifications, phases, parameters, failures, framework, and procedures of the handover in an LTE-A system. In Section 2.6, related studies are summarized in brief, covering the handover decision of various networks (horizontal and vertical). The handover decision evaluation is discussed first, after which the horizontal and vertical handover decision algorithms are categorized based on the respective decision technique. Finally, the concluding remarks and summary of research overview are presented in Section 2.7.

2.2 THE WIRELESS NETWORKS REVOLUTION

The revolution of wireless networks started four decades ago. The first version of commercial implementation, that is, the First Generation (1G) network, was observed in 1980. The 1G network depends on Frequency Division Multiple Access (FDMA) and Frequency Division Duplex (FDD) mode systems but only provides voice call services. The 1G network consists of different systems. For example, in North America, it consists of the Advanced Mobile Phone System (AMPS), whereas in the United Kingdom it consists of the Total Access Communication System (TACS) (Chen et al. 2015a; Dahlman et al. 2007). In 1990, the Second-Generation (2G) mobile network system was commercially deployed. This network is based on Time Division Multiple Access (TDMA) or Code Division Multiple

DOI: 10.1201/9781003648130-2

Access (CDMA) as well as the FDD digital (circuit switch technology) mode. The Global System for Mobile communications (GSM) and North American Interim Standard system (IS-95) commercially implemented the 2G-network (Dahlman et al. 2007).

Later, 3G was developed with the coordination of International Telecommunication Union (ITU) requirements under the name of Universal Mobile Telecommunications System (UMTS). This network uses Wide Code Division Multiplexing Access (WCDMA) technology based on the Direct Sequence Spread Spectrum (DSSS) (Ramachandran 2003). It supports larger bandwidth, capacity, and data rates. The 3G network is known as WCDMA in Europe, CDMA in North America, and Time Division Synchronous Code Division Multiple Access (TD-SCMA) in China. In 2007, the Worldwide Interoperability for Microwave Access (WiMAX) or (IEEE 802.16e) officially became the 4th member of the 3G global standards family (Chen et al. 2015b). LTE and LTE-A adapt the 3G technology and keep up with the standards of GSM, GPRS, and UMTS. When the International Telecommunication Union Radio communication (ITU-R) was launched, it set International Mobile Telecommunications Advanced (IMT-Advanced) high-level requirements. These requirements are a combination of functionality and flexibility to support a huge domain of services and applications, compatibility between heterogeneous networks, compatibility between IMT and other network services, enhancement of roaming capability, and support for advanced services by increasing bandwidth rates (ITU-R 2008) to cover the growth of wireless network demands.

The 3GPP network developed a new Radio Access Technology (RAT) or E-UTRAN, also known as the Fourth Generation (4G) network (3GPP 2010a; Salman et al. 2014). In June 2012, ITU-R formally approved the LTE-A and wireless MAN-advanced (IEEE 802.16m) as part of the IMT-Advanced (4G) standards. The LTE-A and MAN-Advanced systems support both FDD and Time Division Duplex (TDD) modes (Chen et al. 2015b; Kim et al. 2010). LTE-A uses Orthogonal Frequency-Division Multiple Access (OFDMA) in DL and Single-Carrier Frequency Division Multiple Access (SC- FDMA) in UpLink (UL).

For next-generation wireless communications, many researches have started working on the Fifth Generation (5G) network. For instance, the EU Project "Mobile and wireless communication Enablers for the Twenty-Twenty Information Society," METIS 2020, was launched in 2012 to arrive at an international definition for 5G. Moreover, the University of Surrey in the United Kingdom allocated £35 million to fund the 5G invention, which offers enhancement to mobile operations and improves the spectrum efficiency for data rates beyond 4G (Chávez-Santiago et al. 2015). Moreover, Bangerter et al. (2014) mentioned the expectation for 5G technology such as small base stations that already exist and new base stations that will be developed and deployed in the future. As per the author's expectation, macro cells will not continue to be expanded in capacity. Thus, in 5G, the heterogeneous network and new RATs such as the new Wireless Wide Area Network (WWAN) and the Wireless Local Area Network (WLAN) will be used more than other generations. Therefore, there is much expectation surrounding the future of 5G, as illustrated in Figure 2.1.

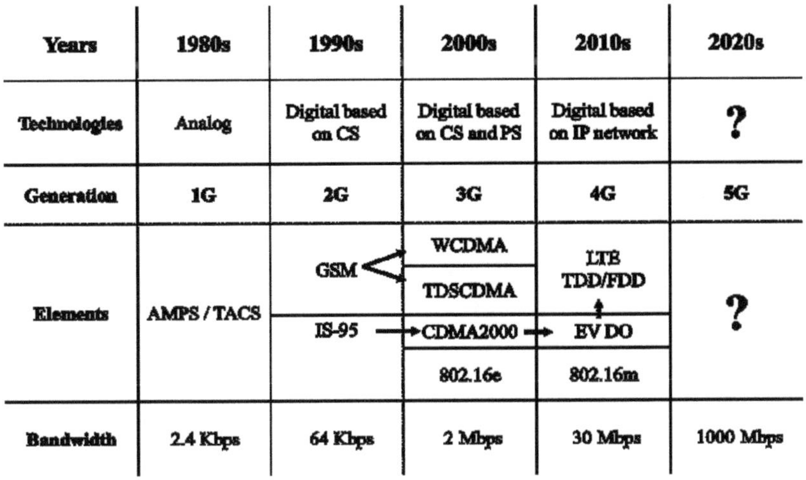

FIGURE 2.1 The evaluation of a wireless network system. (Source: Chen et al. 2015b.)

2.3 LTE-A OVERVIEW

The LTE-A system is viewed as the next main transformative step toward advancing the progress of LTE Rel.8 and Rel.9. These characterize and indicate new attributes and strategies for the next generation of mobile frameworks. The development of LTE Rel.8 and Rel.9 aims to fulfill the requirements of the ITU to become an IMT-Advanced system. The main requirements are (3GPP 2014a; Akyildiz et al. 2010; Tran et al. 2012):

1. Offers higher data rate
2. Provides higher spectral efficiency
3. Supports more extensive transmission bandwidth
4. Provides a flexible spectrum
5. Escalates system capacity, in terms of UE number per cell
6. Supports high mobility velocity
7. Minimizes delay performance of the UE and control planes
8. Supports seamless connectivity and smooth handovers
9. Provides a high QoS

2.3.1 LTE-A Architectures

The LTE/LTE-A system has evolved from being a 3G system to integrating with the IMT-Advance standard that was established by the ITU (ITU-R 2008). Thus, the 3GPP created two components, the RAT and the Core Network (CN), as well as distributed the mobility management between them. In RAT, various technical improvements to the 3G access network were combined with the 3GPP to result in an E-UTRAN in LTE. On the other hand, the CN, which is referred to as the System

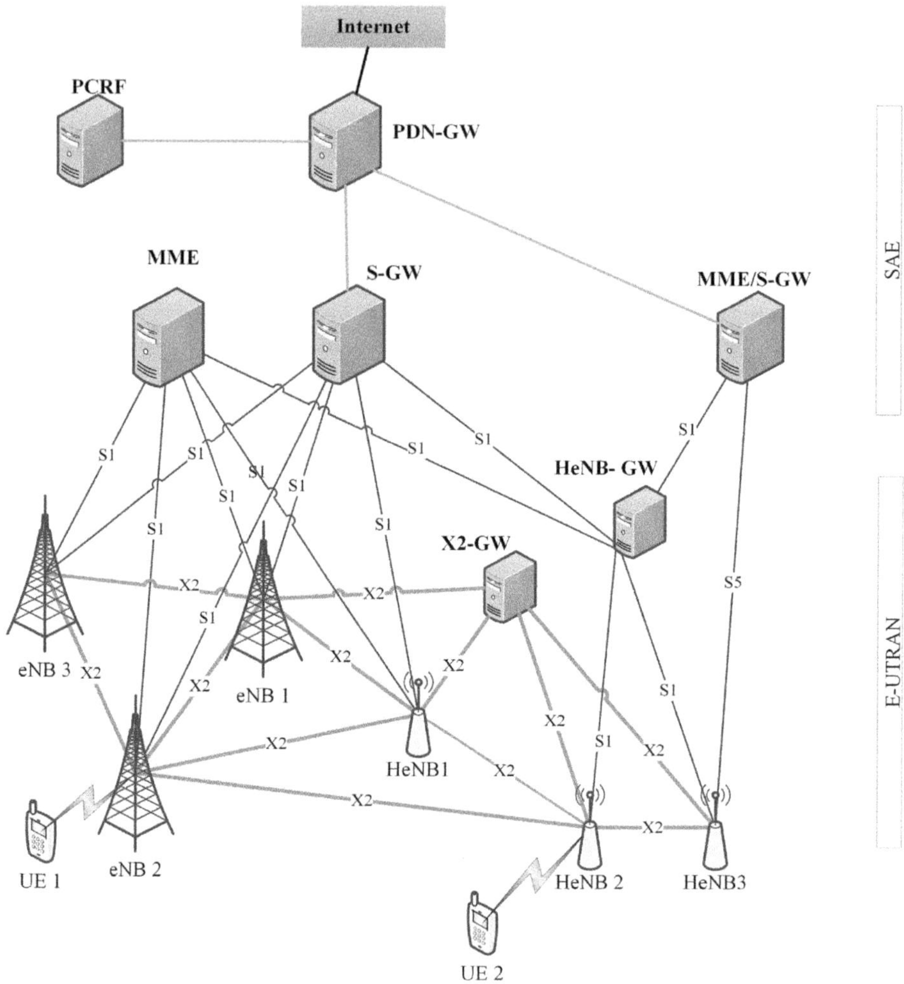

FIGURE 2.2 EPC architecture of LTE-advanced. (Source: 3GPP 2016b.)

Architecture Evolution (SAE), supports compatibility with non-3GPP wireless technology (Salman et al. 2014). Together, E-UTRAN and SAE from the EPC (Akyildiz et al. 2010) are illustrated in Figure 2.2.

2.3.2 The Core Network Control Plane

The Mobility Management Entity (MME) and Serving Gateway (S-GW) are the two entities in EPC, which are involved in supporting the mobility management and control between RAT and CN, as shown in Figure 2.2. The MME is the main control structure in LTE\LTE-A, which is responsible for the mobility management inside the RAT section, bearing management, security, paging management (mobility handling), and admission control. In addition, MME uses the Non-Access-Stratum (NAS) protocol to establish the bearer paths and connection for UEs. The S-GW is proposed for routing and forwarding functions as well as for maintaining the data paths between the cells and Packet Data Network Gateway (PDN-GW). Thus, packet switching will be used in RAT to enhance mobile services, by providing seamless Internet Protocol (IP) connection between UE and PDN-GW, so as to avoid any disruption in the UE applications through their mobility (Akyildiz et al. 2010). In addition, S-GW works as the mobility anchor for both types of handovers for the 3GPP access. The PDN-GW provides connectivity between UE and the external data network. Moreover, the Policy and Charging Rules Function (PCRF) is responsible for service flow policy and charging rules (Akyildiz et al. 2010; Salman et al. 2014; Tran et al. 2012).

Even more, the E-UTRAN network includes the HeNB Gateway (HeNB-GW) entity as logically located and its act as a concentrator for control-plane messages from small cells (HeNBs) and forwards them to MME. hence, HeNB-GW appears to HeNB cells as an MME and to SAE as simple eNB (Damnjanovic et al. 2011). Therefore, a large number of HeNB cells remain scalable in MME. In the event that an X2-Gateway (X2-GW) is available; the HeNB registers with the X2-GW at power on or after any change of transport network layer addresses. The X2-GW is responsible for routing X2 messages to target cells depend on the routing received X2 message, also inform the relevant HeNBs or eNBs which they have the X2 Application Part (X2AP) connection upon detecting the HeNB or eNB signaling the connection to that HeNB or eNB is not available.

2.3.3 The Access Network (E-UTRAN)

In E-UTRAN, the 3GPP uses OFDMA radio interface in order to immune the Inter-Symbol Interference (ISI) from the frequency eclectic fading channels of cellular mobile, especially in the high mobility. In the UL it implemented the SC-FDMA radio interface which is allowed multiple users to assign communication resources speedily (3GPP 2011b; Jimaa & Alfadhl 2011). Based on the OFDMA framework, each radio frame has 10 sub-frames, and each sub-frame contain two slots in the time domain, whereas each slot matches one of the Physical Resource Blocks (PRB) in the frequency domain as illustrated in Figure 2.3.

Moreover, the OFDMA isolates the available bandwidth into multiple narrow-bands and mutually orthogonal sub-carriers, even more; the smallest transmission unit in DL system is the PRB which is involved in a couple of Resource Block (RB). Each PRB dwells 180 KHz bandwidth and contains 12 sub-carriers, each sub-carrier dwells 15 KHz bandwidth (3GPP 2015c). Moreover, each sub-carrier contains 7 resource elements when normal Cyclic Prefix (CP) is utilized, where CP is used as interval guard between the sub-carriers and resource element which is the fundamental unit in LTE (3GPP 2015b). However, the OFDMA consumes a high level of power in the cellular mobile system, thus, SC-FDMA technology is used in UL to provide a more efficient usage of battery in UE (3GPP 2016b).

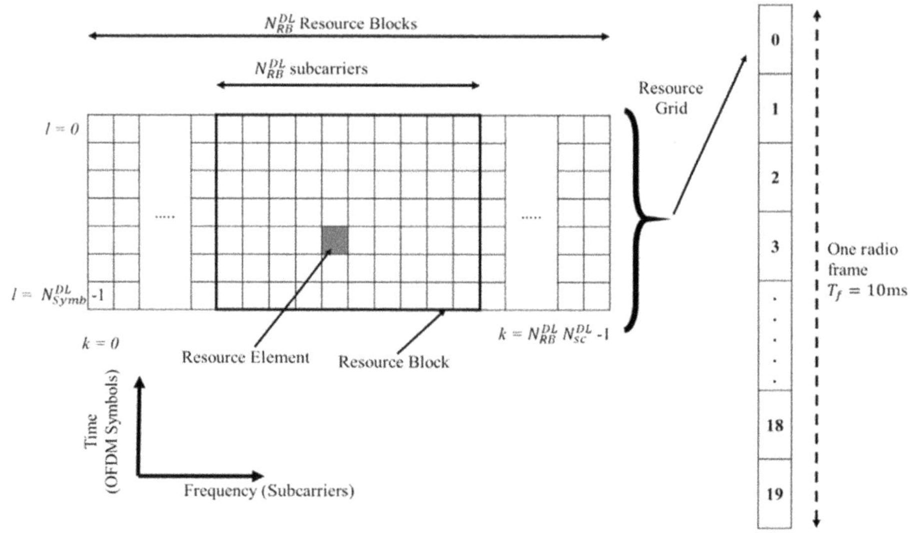

FIGURE 2.3 Downlink resource block in LTE-A system. (Source: 3GPP 2015b.)

In addition, 3GPP improved E-UTRAN through supporting the spectrum flexibility by using the two ways of duplexing, FDD and TDD (Chen et al. 2015b). The FDD mode allows for DL and UL to be transmitted simultaneously in different frequency bands while the TDD mode allows for the DL and UL to be transmitted in the same frequency bands with different timeslots. Thus FDD uses paired spectrum while the TDD used unpaired spectrum. Furthermore, the 3GPP supports six kinds of flexible bandwidths (1.4 MHz, 3 MHz, 5 MHz, 10 MHz, 15 MHz and 20 MHz) as well as it supports the network operator in choosing the transmission bandwidth based on the available bandwidth. In addition, the E-UTRAN technology also supports different frequency bands, which means to be compatible with any system that deployed within 900 MHz, 2.1 GHz and 2.6 GHz spectrums (Jimaa & Alfadhl 2011).

2.3.4 E-UTRAN Control Plane

The main access network (E-UTRAN) component is eNB. In LTE\LTE-A, the eNB wireless access network combined NodeB (wireless access in 3G) with the Radio Network Controller (RNC) to distribute the management and control to each eNB. Each eNB provides controlled operations that terminated towards UEs, such as Radio Resource Management (RRM) which supports the mobility of UE in between eNBs without the user intervention, admission control, RB scheduling of DL and UL to UEs, paging, and handover management (Tran et al. 2012; Volkan et al. 2014). eNBs are connected with each other and with small cells via X2 interface and also connected to CN via S1 interface. The S1 interface is divided to the S1-MME interface used for MME connection and S1-S-GW interface used for the S-GW connection, as illustrated in Figure 2.2.

In the intra-LTE-A access network, the X2 interface is used for handover process. The cells use X2 interface to send and receive messages between them to support UE's mobility. The UE's serving cell sends handover request and the UE's future cell receives the message through X2 interface and reserves resource and sends back Handover Acknowledgment message through X2 interface. The S1 interface is used as the S1 Application Part (S1AP) protocol in radio network signals. The S1 interface handles all actions among CN and E-UTRAN. Moreover, it can carry messages transparently among CN and UE. The S1AP protocol mainly supports E-UTRAN procedures from the CN, transmits transparent non-access signaling, and implements the mobility task (3GPP 2012b).

The HeNB is the second important component of E-UTRAN which is connected to HeNB-GW via the S1 interface. The X2 interface is also used to connect between each other and with eNBs. The HeNBs are connected directly to S-GW via S1 interface. In the 3GPP Release 11, the 3GPP was proposed to integrate security gateway between HeNB and HeNB-GW for mutual authentication for UE data while being on the IP network (3GPP 2012b). Moreover, in HeNB there are three types of accessibility: the open access, the closed access, and the hybrid access.

2.3.5 LTE-A E-UTRAN New Features

In LTE-A, additional E-UTRAN features are supported compared to LTE as follows:

1. Carrier Aggregation (CA): attempts to raise the data rates to 1Gbps for DL and to 500 Mbps for UL, by enabling bandwidth extension to the limit of 100 MHz by aggregating five carriers into either contiguous or noncontiguous component carriers. Each carrier has 20 MHz (Kanchi et al. 2013; Shen et al. 2012b; Tran et al. 2012).
2. Advanced Multiple Inputs Multiple Outputs (MIMO): in an attempt to raise the requirements of channel transmission efficiency more than those in LTE, 3GPP increased the bit rate in different data stream transmissions by supporting the configuration of four antennas in UL and to eight antennas in DL (Hätty et al. 2016; Tran et al. 2012).
3. Coordinated MultiPoint (CoMP): in an attempt to reduce the interference with UEs which are located in multi-distributed network cells, the cells cooperate with each other to improve the system and/or cell edge throughputs by reducing the inter-cell interference in both DL and UL (Akyildiz et al. 2010; Tran et al. 2012; Lee et al. 2012) .
4. Relay Node (RN): it is used to improve in terms of coverage and throughput of the cellular network. RN enhances the coverage and throughput in targeted area inside eNB without the need of wired backhaul connection through signals relaying in the small entity (Chen et al. 2015a). In addition, the cost and complexity of wireless relays are less than eNBs; thus, RNs are used to

support the group mobility and capacity at the edges of eNB (Akyildiz et al. 2010; Tran et al. 2012; Yang & Winter 2015; Damnjanovic et al. 2011).

5. Heterogeneous wireless Networks (HetNet): they are used to increase the coverage and capacity within the umbrella of the eNB cells. The structure of LTE-A E-UTRAN supports several low-powered small cells such as a microcell, picocell, and femtocell (H-eNB). This type of wireless networks is called HetNets (3GPP 2010a, 2013).

All these enhancements will be more efficiently achieved in the line of LTE-A system development, as illustrated in Table 2.1. The table summarizes the evolution of LTE and LTE-A system starting from Rel.8 to Rel.13 and beyond.

2.4 HETEROGENEOUS WIRELESS NETWORKS

In homogeneous networks, base station deployment is based on cells of a similar type, similar transmitting power, and similar backhaul connectivity. These cells are deployed to cover the user's demand in a wide coverage area and so to minimize the inter-cell interference. In recent years, data demand in the wireless communication is sharply increased; therefore, the addition of more carriers is needed. Densifying the eNB cells can enhance the network capacity but the deployment process of eNBs is costly and complex, especially in the urban areas. To cope with this problem, a new deployment system is needed. The vertical wireless network, which is known as (HetNet), is considered to be a future solution for facing the increasing demands for wireless communication (Naeem et al. 2016), as illustrated in Figure 2.4.

HetNets bring a different approach to the standard cells deployment system. In HetNets, a low powered node can be deployed within the eNBs coverage area. The small cells deployment has a better adoption for expanding the data traffic demands, providing a uniform user experience, and supplementing the eNBs networks. Micro, Pico, Femto (HeNB), and RN cells are the most feasible and efficient strategy. Micro and Pico cells enhance the coverage and the capacity of wireless networks in the places where the traffic data demand is high. The RN cells are deployed in the places where the backhaul wireline is not reachable.

2.4.1 Femtocell

The HeNB cells provide better service indoors and can be deployed even by enterprises or users (unplanned) to handle the traffic for a Closed Subscriber Group (CSG) (Xenakis et al. 2014). The HeNB cells are low-powered nodes, which are prepared with omnidirectional antennas with a small coverage range (30–40 m). Moreover, HeNBs can connect to the CN within the user's broadband connection. Due to the small coverage range, a huge number of HeNBs is deployed in eNBs. Also, the number of users that each HeNB supports is limited and grouped into three groups according to HeNB constraint policies. These groups include open access HeNBs, which support access to all users; closed access HeNBs, which support the access to

TABLE 2.1
Comparison between LTE and LTE-A systems

	Rel.8	Rel.9	Rel.10	Rel.11	Rel.12	Rel.13
Technology	LTE	LTE	LTE-A	LTE-A	LTE-A	LTE-A
Last Frozen Date	December 2008	December 2009	March 2011	September 2012	June 2014	June 2016
Support CA technique	No	No	Yes	Yes	Yes	Yes
Peak Data Rate DL	10	10	10	15	1	1
(Mbps) UL	5	5	5	7.5	1	1
MIMO DL	4x4 MIMO	4x4 MIMO	up to 8x8 MIMO	up to 8x8 MIMO	up to 8x8 MIMO	up to 8x8 MIMO
UL	SISO	SISO	up to 4x4 MIMO	up to 4x4 MIMO	up to 4x4 MIMO	up to 4x4 MIMO
Duplex mode	Full	Full	Full	Full	Half	Half Low power
UE transmit power (dBm)	23	23	23	23	23	20
UE received bandwidth (MHz)	20	20	20	20	20	1.4

Source: 3GPP (2014b); 3GPP 2014c; 3GPP 2014d; Ghosh et al. 2010; 3GPP 2016c

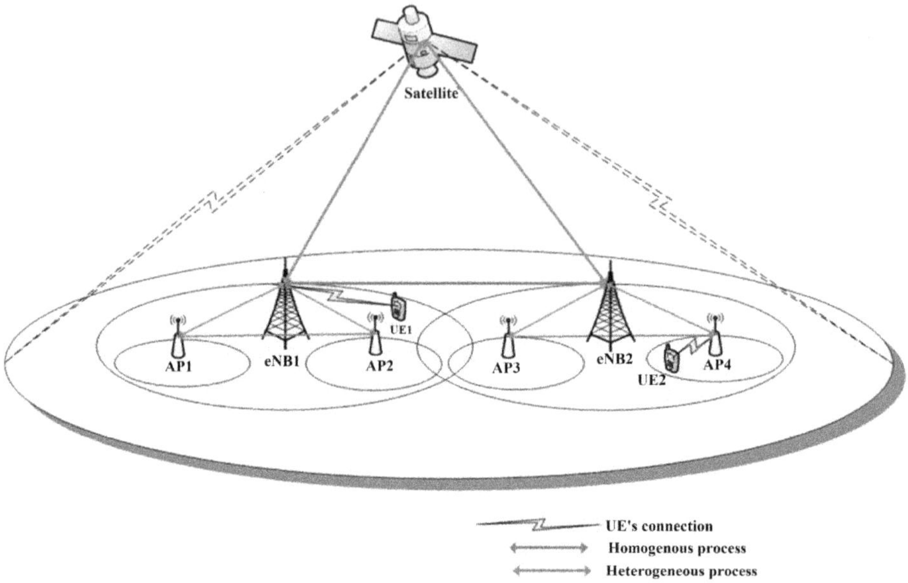

FIGURE 2.4 Both homogeneous and heterogeneous networks.

users in the CSG group; and hybrid access HeNBs, which support access to all users with priority given to users in CSG.

2.4.2 Challenges of HetNets

In spite of the significance of using HetNets to raise wireless network throughput, HetNets could also lead to many technical challenges. These challenges are listed in the following.

2.4.2.1 Inter-Cell Interference

Minimizing cell size means adding more small cells to a particular area to guarantee similar coverage. In this way, the user may see more different and more grounded out-cell interference. Moreover, the concurrence of powerful eNB cells additionally complicates the issue by presenting cross-tier interference among the eNB cell and small cells. However, an efficient interference management solution is needed for the deployment of HetNet (Xenakis et al. 2014, 2016).

2.4.2.2 Resource Sharing for Unplanned Deployment

Small cell deployment is usually based on demand and often not planned carefully. This means that frequency resources could be restricted and rendering the frequency division of small cell deployment is not viable. Thus, the resource sharing procedure needs to be designed carefully in order to adapt dynamically to various network topologies (Li et al. 2016).

2.4.2.3 Mobility management

The primary critical issue in deploying intensive small cells is the increasing handover number, which calls for advanced mobility management. In addition, the irregularity of DL and UL coverage poses more challenges in addressing DL and UL handovers (Gódor et al. 2015).

2.5 HANDOVER OVERVIEW

In wireless network systems, handover is the major of RRM process that boosts reliable UE connectivity at various mobility conditions. Handover generally keeps up the radio link connection for the UE to the best serving base station in coverage scope. Handover procedure, also called handoff, can be defined as the major strategy which permits the UE that is connected with serving base station to exchange to the neighbor-based station (target) without session disconnection (Xenakis et al. 2014, 2016). Therefore, the real-time applications strongly depend on the seamless handover. In E-UTRAN wireless system, the RSS is used to maintain the handover process when the UE moves within the cells. The UE performs the handover process when the neighbor cell provides better RSS. The handover procedure scenario is illustrated in Figure 2.5.

Furthermore, this scenario of handover process is used in both horizontal and vertical types of handover procedure in LTE-A. In horizontal handover, the UE transfers from one network cell to another network cell within the same wireless technology, while in a vertical handover the UE transfers from one network cell to another network cell with different wireless technologies.

Moreover, two main types of vertical handover interoperability are supported in LTE-A wireless system. The first one is among 3GPP that is variated in RAT, such

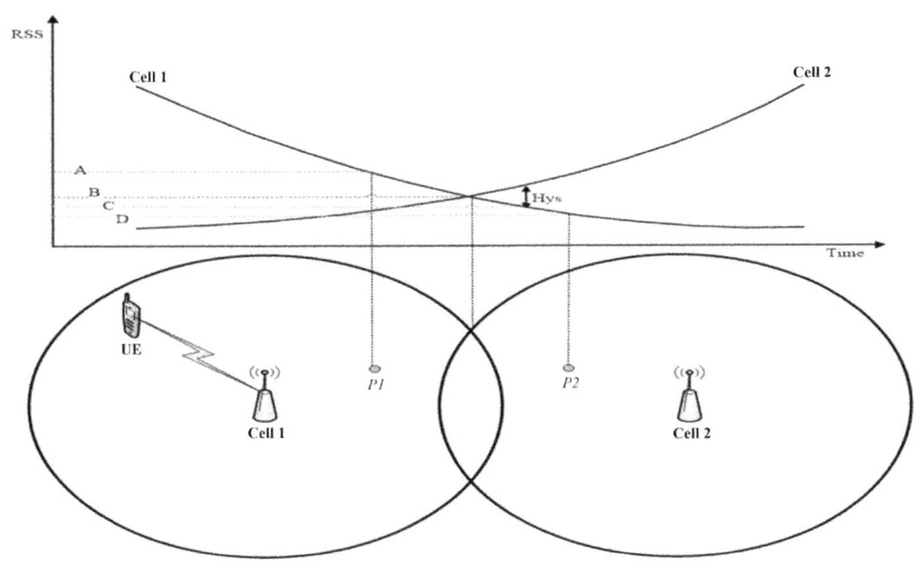

FIGURE 2.5 Handover procedure scenario.

as between eNB and HeNB, and the second one is among non-3GPP wireless technology, such as from WiMAX to LTE-A or vice versa (Chandavarkar & Guddeti 2016; Gódor et al. 2015; Márquez-Barja et al. 2011).

2.5.1 Handover Techniques

In wireless networks, there are two types of handover techniques: the soft handover and HHO (3GPP 2007; 3GPP 2014g). Both techniques are discussed in the following subsections:

2.5.1.1 Soft Handover

The soft handover is a type of handover technique that is also called the connect-before-break process. A new UE radio link is established with the target cell while an old radio link to the serving cell is maintained. Thus, the UE keeps connected to at least one radio link in the network system. Additionally, the UE can also obtain data services from several cells simultaneously (3GPP 2011a). The soft handover was introduced in the UTRAN structure of 3G wireless networks. The RNC is a centralized device that carries out the handover management of each UE in the wireless network, as illustrated in Figure 2.6.

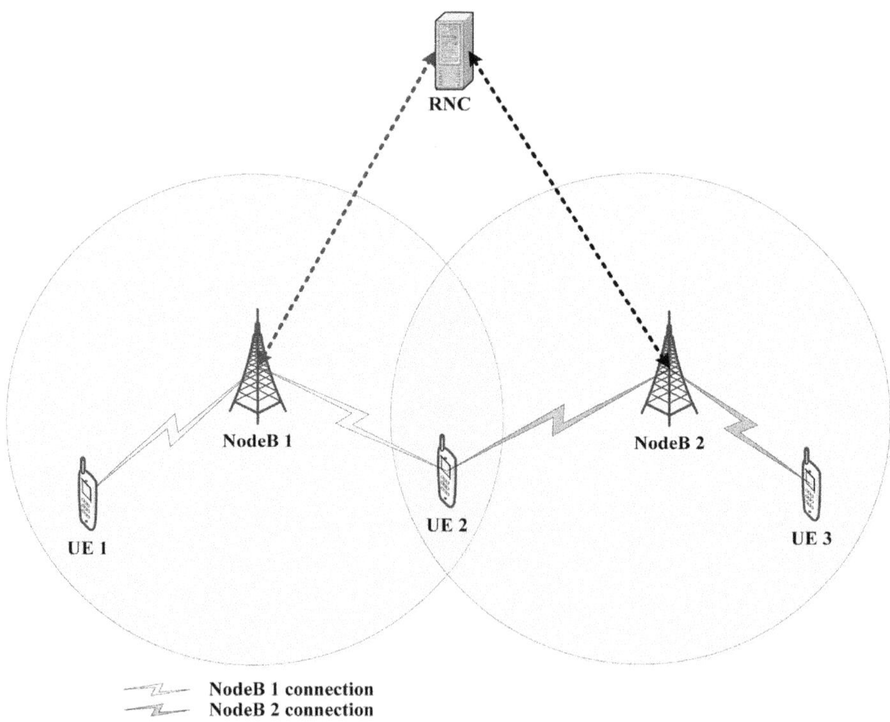

FIGURE 2.6 The soft handover technique.

Based on Figure 2.6, when the user reaches the overlapping area, which is between NodeB1 and NodeB2, the UE will be connected to both NodeB cells. Despite the advantages of the soft handover in maintaining an active session, it is not as suitable as HHO for real-time applications such as voice and packet call drops. Furthermore, the soft handover requires more signaling measures in WCDMA (Chang et al. 2009).

2.5.1.2 Hard Handover

HHO is also known as the break-before-connect process. The procedure of HHO is executed by allowing the UE to break a present connection with the serving cell before making a new connection with the target cell (Lin et al. 2005). Consequently, the UE communicates with only one cell at each time slot during the HHO process. After releasing the radio link of the serving cell, a new radio link will be initiated with the target cell. Figure 2.7 illustrates the HHO technique when the UE moves from cell 1 to cell 2 and only has one radio link connection. In an LTE-A E-UTRAN system, the HHO is adopted due to the flat IP-based structural design, which addresses the shortage in the soft handover. The HHO in LTE-A is initiated when the RSS of the UE target cell is higher than the RSS of the serving cell plus the Hys value in the time to trigger (TTT) interval. Moreover, regarding RNC, which is in the E-UTRAN system, the handover decisions are distributed to eNBs without the intervention of MME/S-GW instead of being in centralized controllers as per the soft handover (Sipila et al. 1999). Also, the X2 interface is used to exchange messages between inter-eNBs.

Moreover, the interruption in handover time, data loss, and buffering will be in the UE plane, and data forwarding would be within distributed controllers in the handover. There are some malfunctions of HHO, such as high data loss, carrier interferences,

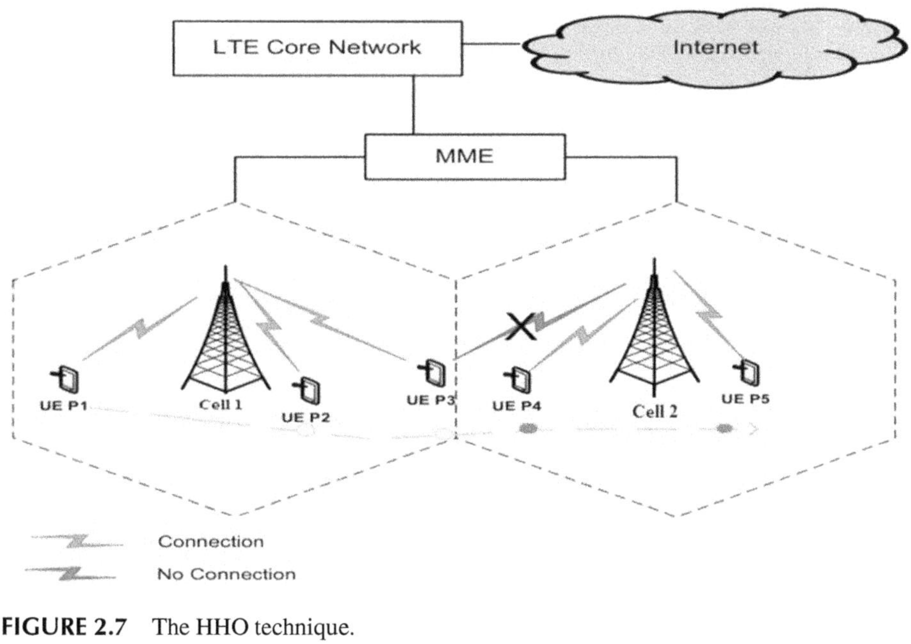

FIGURE 2.7 The HHO technique.

and difficulty in maintaining QoS requirements due to the delay in handover through eNB migration (Chang et al. 2009). Nevertheless, HHO is still fast and reliable for high-velocity handovers and because it is beneficial for UEs as it reduces the complexity of the network.

2.5.2 Horizontal and Vertical Handover

Several wireless network technologies are present recently; two main types of handover procedures are supported: horizontal and vertical handovers (Damnjanovic et al. 2011; Gódor et al. 2015; Márquez-Barja et al. 2011). The handover procedure in the horizontal handover is executed among two network cells; both of them belong to the same type of RAT such as between eNB and eNB, as shown in Figure 2.8 (a). On the other hand, the vertical procedure is executed among two network cells which belong to different RAT. In LTE-A vertical handover, there are two types of vertical handover: the first one is among 3GPP wireless technology, such as between eNB and HeNB, and the second one is among non-3GPP wireless technology, such as between eNB and WiFi or vice versa. Figure 2.8 (b) and (c) illustrates both types of vertical handover when UE moves in different RATs. Since this study is concentrated on handover performance in the LTE-A system, the horizontal and vertical handovers between eNBs and HeNBs are considered.

The vertical handover will have a wider use in the future and will execute handover between various RATs. Thus, 3GPP has a standardized handover process between LTE-A and UTMS in 3GPP (2014f). Standardization of the handover process between LTE-A and WLAN is not available yet, while 3GPP in the 3GPP (2014e) suggested adding a new element to the core network which is called Access Network Discovery and Selection Function (ANDSF). The aim of using ANDSF function is to maintain UEs connection with non-3GPP RATs.

In the vertical handover that is between eNB and indoor cells (HeNBs), there are three handover scenarios:

1. Hand-in: this scenario of handover occurs when the UE moves from the eNB to HeNB coverage area, as the UE has to choose the best HeNB out of a large number of HeNBs.
2. Inter-HeNBs: this scenario of handover is similar to the hand-in procedure, but different in which the UE selects the best HeNB for handover from a large number of HeNBs.
3. Hand-out: this scenario of handover occurs when the UE moves from the HeNB to eNB coverage area. In this case, the handover scenario is not too complex compared to the hand-in because there is only one target eNB for the UE to handoff to.

2.5.3 LTE-A Handover Challenges

The current LTE-A features of both types of the wireless networks provide enhancement in coverage, capacity, and data rates compared to the previous wireless network

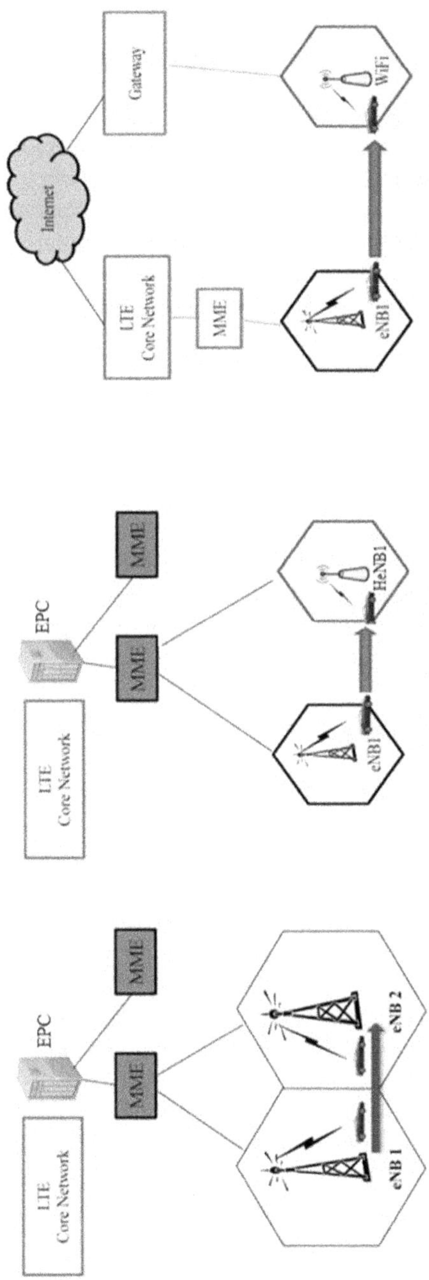

FIGURE 2.8 Horizontal and vertical handovers: (a) horizontal handover, (b) vertical handover among 3GPP RAT, and (c) vertical handover among non-3GPP RAT.

(Akyildiz et al. 2010). Moreover, using of HHO procedure in LTE-A provides more benefits, such as seamlessness, reliability, and reducing the complexity in wireless networks. However, the HHO still causes packets drop. In addition, because of the density of small cells deployment and high velocity of UEs, a number of challenges influence the handover decision scheme. These challenges are concluded as follows:

1. High mobility velocity nature of UEs is the main challenge that is faced in selecting the proper target network cell to serve the UE for a long time. In two-tier networks, the redundant handover will be high when the UE moves fast.
2. The interference is also another challenge; adding more number of small cells is a solution for raising the capacity but then the interference increases and mitigates the handover decision efficiency. Therefore, an interference management or coordination is required; interference mitigation is considered a hot topic among researches. In Lopez-Perez et al. (2011), the authors introduced an enhancement to the inter-cell interference coordination in a wireless network and in 3GPP (2012d), the authors introduced a simulation of the enhancement of inter-cell interference coordination and proved its enhancement with regard to HF and handover link failure.
3. The variation in transmitting power between the two types of wireless networks is another issue; when the HeNB location is near to eNB tower, the RSS of the HeNB is always less than the RSS of the eNB cell.
4. The optimization of handover decision parameters threshold values is the next challenge. These parameter values should be dynamic and adapted to radio access network and UE cases, respectively.

2.5.4 LTE-A Handover Phases

Each handover process in LTE\LTE-A consists of three phases: the handover measurements, the handover decision, and the handover execution (Anas et al. 2007).

2.5.4.1 Measurements Phase

In this phase (also called information gathering), the UE periodically obtains RSS values during an active mode session from serving cell and neighbor cells to decide which neighbor cell has a better signal condition for UE switching in handover decision time (Kurjenniemi & Henttonen 2008). The continuous scanning measurement serves the target cell selection so that it keeps the QoS and the availability of network indicators (Miyim et al. 2014). In Han and Wu (2010), the authors mentioned that many measurement factors have affected handover decision. Some of these factors are related to network performance and control, such as Reference Signal Received Power (RSRP), RSQ, Signal Interference plus Noise Ratio (SINR), Received Signal Strength Indicator (RSSI), Signal Noise Ratio (SNR), and Carrier Interference Ratio (CIR). On the other hand, other factors are related to UE, such as velocity and battery life.

The measurement event is triggered when the RSS measured values that are obtained by UE are met with the handover event condition (A3 event), and then it reports to the serving cell. A3 event is an E-UTRAN measurement reporting event criteria (3GPP 2012a) . Furthermore, the UE triggers A3 event when the QoS is being less than the threshold value and the RSS of the serving cell starts to be weaker than its neighbor cells or when there is a limitation of channels availability which is a result of congestion. In Cheng-Chung Lin (2013), it was mentioned that the measurement phase starts from UE through receiving three input and sending three output parameters to the serving cell to make a handover decision.

The input parameters are the priority of applicable frequencies of RAT, the neighbor cells cases, and the distance between UE and serving cell. The output parameters per contra are SINR, availability cells, and the priority of applicable frequency of RAT. While the neighbor cells list scheduling is required for each measurement report of each UE, the measurement gap time extends and the QoS reduces, thus reduced the number of neighbor cells is mandatory. Therefore, in 2012 3GPP developed an Automatic Neighbor Relation (ANR) standardization scheme (3GPP 2012d) which permits eNBs to maintain neighboring cells list depending on the UE measurement results and neighbor cells operation settings. This list has the neighbor cells identity (cells_id) and the frequencies. In addition, several studies, as in Shen et al. (2012a), Wanalertlak et al. (2011); Watanabe et al. 2011), improved the handover management by reducing neighbor cells list.

2.5.4.2 Decision Phase

After the serving cell decides the handover initialization by using data measurement, the handover decision phase comes next. During this phase, the candidate of neighbor cells will be evaluated and the final result leads to selecting the target cell (Miyim et al. 2014). According to 3GPP (2012a), the serving cell sends handover request message to the target cell that has highest reference signal received power (RSRP) of neighbor cells and Equation (2.1) is achieved as long as the UE has stayed for at least 1 s in the serving cell.

$$Q_n > Q_s + Off_{(s,n)} + H_{ys_s} \tag{2.1}$$

where Q_n is the RSRP of the neighbor cell, Q_s is the RSRP of the serving cell, $Of f_{(s,n)}$ is the individual offset of neighboring cell, and Hys_s is the hysteresis or handover margin parameter of serving cell. Therefore, the hysteresis parameter should set up before handover decision is triggered (Zhu & Kwak 2007). The effectiveness of handover decision is based on right time and precise measurements of handover decision parameters to maintain data connection when UE is being transferred from serving cell to target cell. Thus, in vertical network the density of HeNBs in a particular area makes new challenges, such as the number of signaling measurements and the number of unnecessary handovers based on the small coverage of HeNBs. Moreover, in urban areas, the RSS of eNBs deployment is fairly good and the UE will not measure the existing HeNBs because the signal strength of serving eNB is still desirable (Chang et al. 2012). However, in vertical handover decision the conventional handover procedure needs updating by using additional input parameters with RSS, in order to improve the performance of handover decision.

After requesting handover from the target cell, target cell has its own right to agree or deny access. When requested access is denied, handover failure (HF) occurs and sequent HFs lead to radio link failure (RLF) (Suleiman et al. 2016). On the other hand, if the requested access is accepted, then the target cell starts to prepare itself for the incoming UE and handover execution phase starts.

2.5.4.3 Execution Phase

Execution phase is related to signaling and transferring of radio path procedures to perform the mobility of the connected UE. The data sending process is started from the serving cell to the target cell, and UE also needs to set up a correlation with target cell through different operations, such as synchronization, authentication, and network reconfiguration. Finally, the UE sends a completion message to the target cell to indicate that the handover is complete (Zhou & Ai 2014).

2.5.5 Handover Parameters

A handover procedure in LTE\LTE-A uses different parameters which control the performance of handover process, especially in triggering point. These parameters need a setting to get optimal values which affect handover decision as shown in Figure 2.9, where P_1 and P_2 are the received signal strength of serving and target cells, respectively.

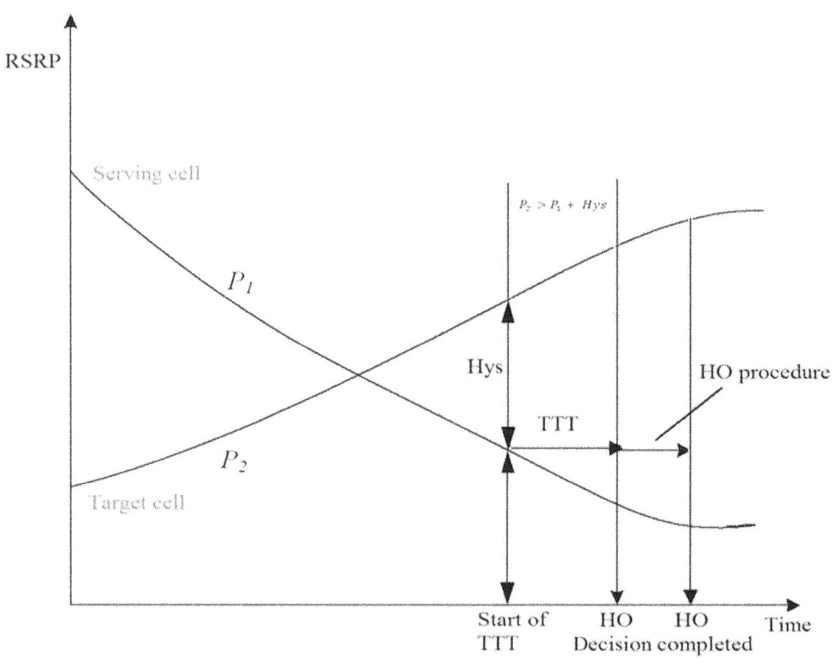

FIGURE 2.9 Handover parameters in LTE\LTE-A. (Source: Fei and Fan 2012.)

2.5.5.1 RSRP and RSQ Parameters

RSRP refers to the strength of signals transmitted by an LTE\LTE-A specific cell, and these signal values use a synonymous path loss in between UE and the different candidate cells. The RSQ refers to the quality of a specific cell signal, which is calculated by finding the ratio of RSRP of the specific cell (target cell) to the total of the received interference power (which is indicated by RSSI) at the UE site, which involves estimating the SINR of the target cell (3GPP 2011b). Both RSRP and RSQ metrics belong to the DL measurement channel, which is made on the UE site and forwarded to the serving cell as a measurement report message when the handover event occurs. Thus, the variation in RSRP and RSQ threshold values results in different trade-offs and enhances the performance during the initiation phase of the handover (Isa et al. 2015).

2.5.5.2 Hysteresis Parameter

The hysteresis parameter, sometimes called the handover margin, controls the handover algorithm by making the signal values of neighbor cells weaker than the real signal values so that they will be strong enough for the UE handover initial decision. Moreover, the hysteresis value is used to prevent the occurrence of a ping-pong handover (this will be explained in handover failures) (Isa et al. 2015).

2.5.5.3 Time to Trigger Parameter

The initial handover trigger depends on the threshold time interval referred to as TTT, for which its threshold value should be included in configurations of measurement. The elongation of the TTT threshold value may reduce unnecessary handover numbers and prevent ping-pong handovers, but will raise the probability of handover failure and cause a delay in the handover. On the contrary, the shortening of the TTT threshold value would result in an early handover. Thus, the optimal TTT threshold value still remains one of the important handover challenges in this field (Kim et al. 2014).

2.5.6 Handover Failures

The handover procedure depends on the handover parameter threshold value. In some cases, the configuration of these parameters is unsuitable and results in HFs; the sequence of HFs leads to RLFs. Therefore, these failures force the UE to reconnect with the serving cell or with the nearest neighbor cell and restart the handover procedure again. In this subsection, the topics of handover failures are defined.

2.5.6.1 Too Early and Too Late Handover

These types of failures occur when the handover procedure is triggered at the incorrect time, either too early (low value of TTT) or too late (high value of TTT). The selection of target cells will be ineffective during the handover decision algorithm (Davaasambuu et al. 2015). Figure 2.10 illustrates these two cases.

2.5.6.2 Wrong Target Handover

This type of failure occurs when the UE exists on the edge of the serving cell and is within the signals of the overlapping neighbor cells; at that time, the UE will try

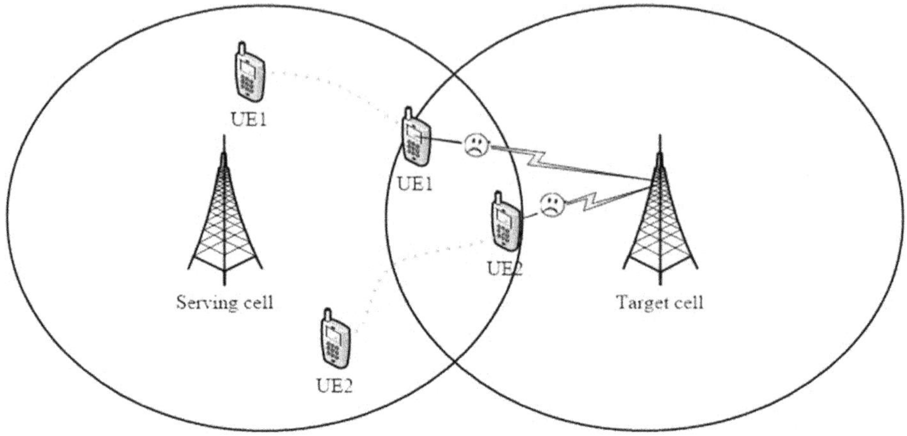

FIGURE 2.10 A too-early handover in UE1 and a too-late handover in UE2. (Source: Davaasambuu et al. 2015.)

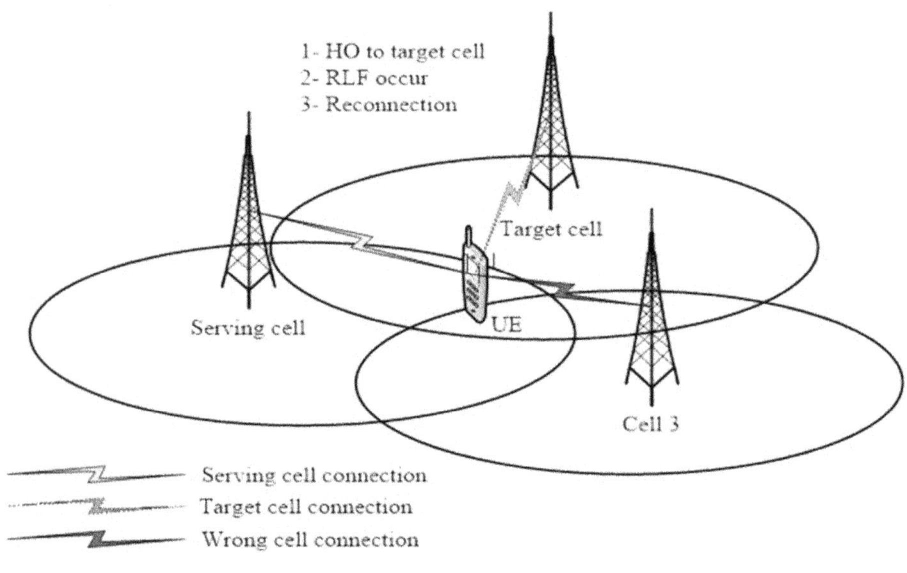

FIGURE 2.11 Wrong target handover. (Source: Wang et al. 2014b.)

to connect with the target cell. During this connection attempt, HF will occur in the target cell in which the UE will try to re-establish a connection with another cell. After establishing the connection, the third cell will inform the target cell through sending an RLF message. Then, the target cell will inform the serving cell via a handover report message, stating that the selection of the target cell was wrong (Wang et al. 2014b). Figure 2.11 illustrates this case more clearly.

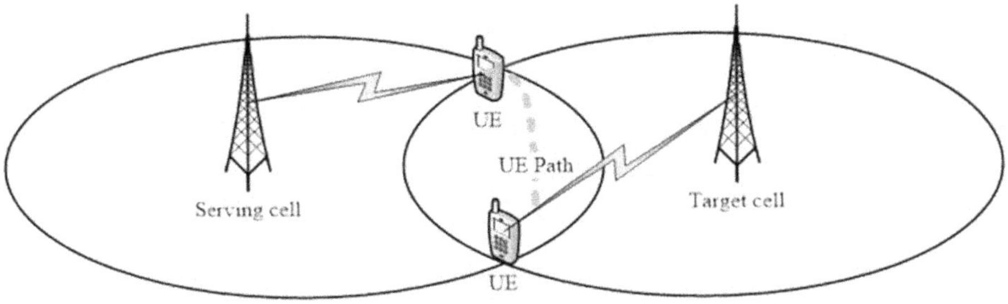

FIGURE 2.12 Ping-pong handover. (Source: Ghanem et al. 2012.)

2.5.6.3 Ping-pong Handover

This type of failure occurs when the UE is located on the edge of the serving cell, thus intersecting with the edge of the target cell in the overlapped area and vice versa. In this situation, the UE hands off to the target cell and stays in the target cell for less than 5 s and then returns back to the previous cell (Li & Wang 2013). Figure 2.12 illustrates the ping-pong handover in more detail.

2.5.7 LTE-A Handover Procedure

Because LTE\LTE-A is supporting various ranges of RAT which are together compatible with non-3GPP networks, three handover signaling procedures are supported based on the interface connection type (without X2, X2, or S1) (3GPP 2013), as explained below.

2.5.7.1 Handover Procedure without X2 Interface

According to this type of handover, cell search, transmitting signaling, and handover decision are depicted in Figure 2.13.

Based on Figure 2.13, the handover procedure consists of 16 steps as explained in the following steps (3GPP 2016b):

1. The source eNB (serving eNB) configures the UE with proximity indication control.
2. The UE detects the nearby HeNB whose CSG-Id is in UE list and then sends proximity indication to its source eNB through autonomous search procedures.
3. The source eNB reconfigures the UE through providing measurement gaps to assist the UE in reporting RAT and frequency. Moreover, the proximity indication may be used by the network to reduce the call for handover preparation of CSG HeNBs when the UE is out of the geographical area of CSG HeNBs list.
4. The UE sends a measurement report message, which contains Physical Cell Identifier (PCI), to source eNB during the handover procedure time. The UE measures the signal strength of the target HeNB.

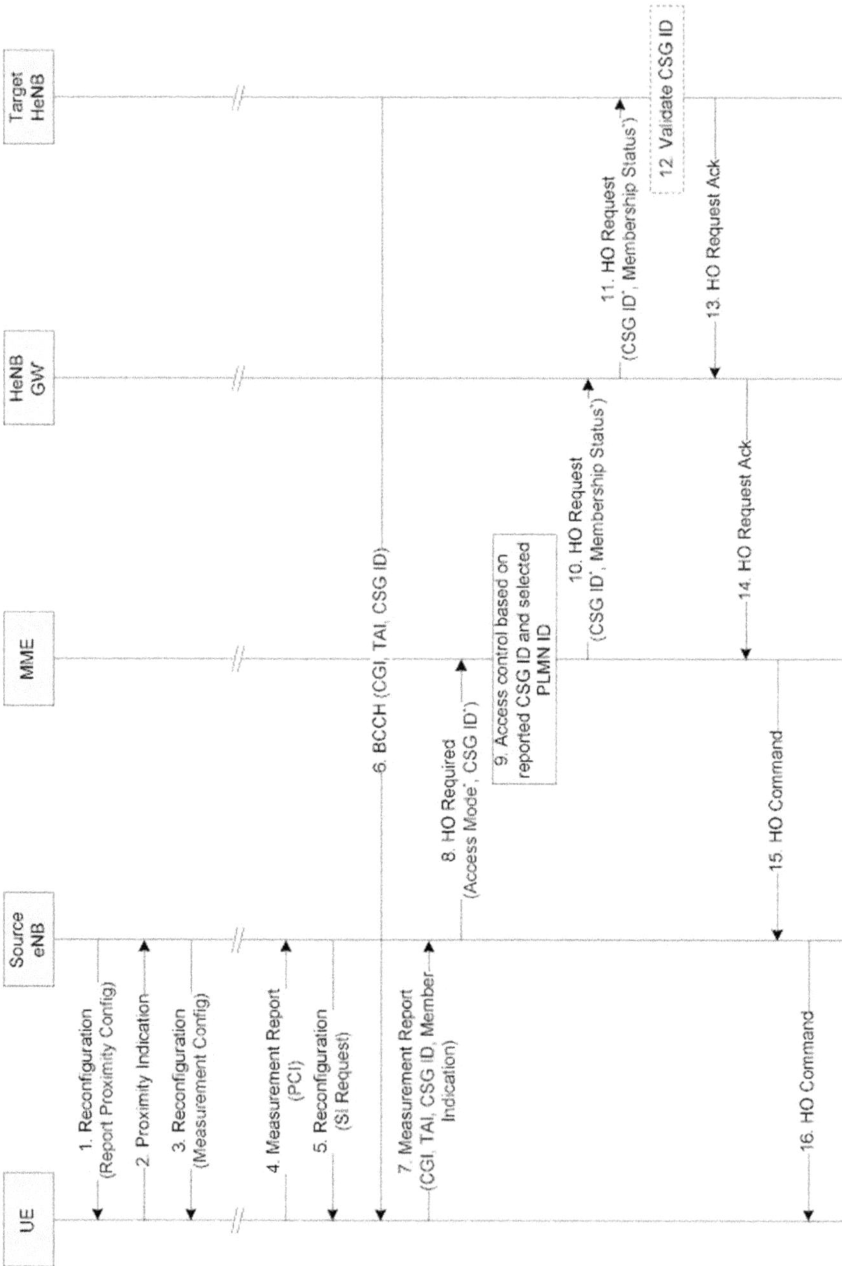

FIGURE 2.13 The handover procedure without X2 interface. (Source: 3GPP 2016b.)

5. The source eNB reconfigures UE to obtain System Information (SI) of a particular PCI (target HeNB) by overhearing the Broadcast Control Channel (BCCH).
6. The UE executes SI obtained by using autonomous gaps, at the end of the autonomous gaps the UE tunes back to the source eNB BCCH frequency.
7. The UE sends the measurement report message, which contains SI acquisition such as Cell Global Identifier (CGI), CSG ID, and extra parameters, to the source eNB.
8. The source eNB encapsulates the data (CGI, CSG ID, and cell access mode) that was sent from the UE by the measurement report in a handover-required-message and then forwards it to MME.
9. The MME implements the UE access control on CSG cell (depending on the CSG ID and the selected target Public Land Mobile Network (PLMN)).If the access control function fails, the MME terminates the handover procedure by responding to a handover preparation failure message.
10. The MME forwards the handover-request-message to the HeNB-GW based on the CSG ID acknowledged in the handover-required-message.
11. The HeNB-GW forwards the handover-request-message to the target HeNB based on the CSG ID acknowledged in the handover-required-message.
12. The target HeNB investigates the CSG ID and allocates the resources for the new UE if it is verified.
13. Check if the previous investigation is verified.
14. The target HeNB forwards the handover-request-acknowledgment to the MME through an HeNB-GW (if HeNB-GW presents).
15. When the MME received the acknowledgment message, the MME sends handover-command message to the source eNB.
16. The source eNB commands the UE to trigger the handover by using the handover command (Radio Resource Control (RRC) Connection-Reconfiguration message including mobility control information).

2.5.7.2 X2-Based Handover Procedure

According to this type of handover procedure, X2 interface is used to direct messaging exchange among the same type of wireless networks, such as inter-eNBs or inter-HeNBs (3GPP 2013), as illustrated in Figure 2.14.
The X2-based handover consists of three phases respectively, as explained below:

i. **Handover preparation phase**
In this phase, the first six steps of the handover procedure are addressed and the measurements messaging are exchanged between the UE and serving eNB periodically to gather information for the handover decision in step 1s. When the measurement report is sent to the serving eNB by UE, and when the measurement parameters are met with the handover triggering criteria, the handover decision will be triggered in step 2. In step 3, the serving eNB sends a Handover-Request (HO-Request) message to the target eNB via an

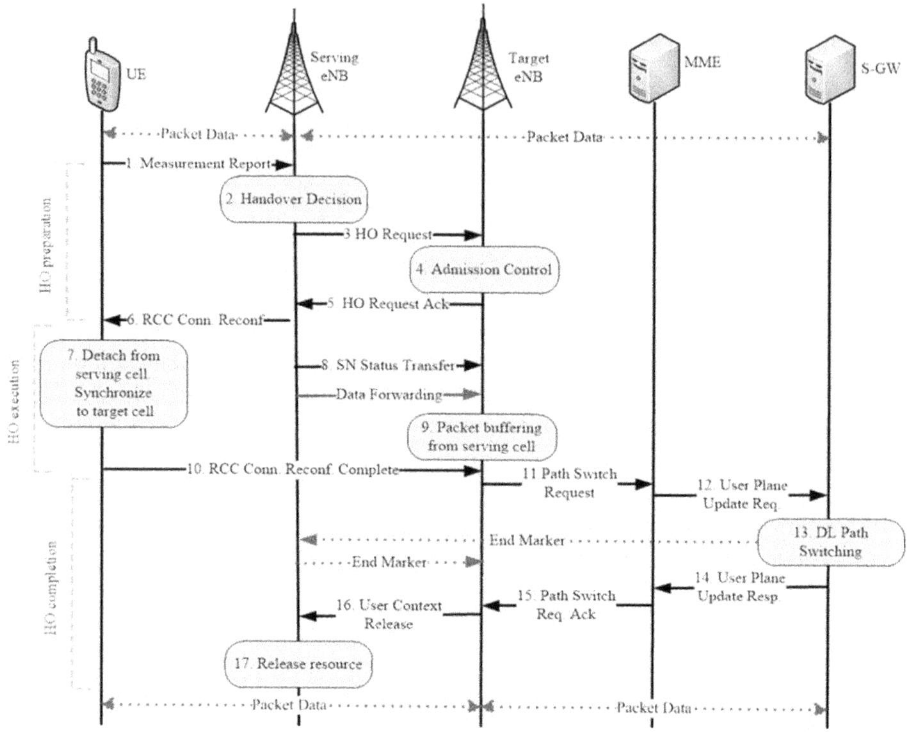

FIGURE 2.14 X2-based handover procedure. (Source: 3GPP 2016b.)

X2 interface. The target eNB performs admission control depending on the serving eNB message to increase the probability of handover success. After that, the target eNB has the privilege to accept or reject the serving eNB handover request. Henceforth, the target eNB will reply back via an acknowledged message (HO-Request-ACK) to the serving eNB through the X2 interface. Hence, the serving eNB will receive the message and start forwarding the RCC Reconn. Reconf. to the UE to request the handover action, whereas the RCC Reconn. Reconf. is the configuration message, which indicates the UE, on one hand, to detach from the serving eNB, and on the other hand to synchronize and attach to the target eNB (Gódor et al. 2015).

ii. **Handover execution phase**

This phase becomes active through steps 7–10 in the handover procedure. The UE detaches from the serving eNB and synchronizes with the target eNB, while the serving eNB sends the Sequence Number (SN) message, which carries the sequence number of transferred message and forwards the data packets that have newly arrived from the S-GW to target eNB via the X2 interface instead of the UE (at this time, the UE is not connected to the network). After the connection is established between the UE and the target eNB, the

UE will start to transmit the RRC Reconn. Reconf. and send the complete handover message to the target eNB.

iii. **Handover completion phase**

This phase covers steps 11–17 in the handover procedure. The target eNB sends a (Path-Switch-Request) message to the S-GW through an MME entity to terminate the data forwarding and to complete the handover. As part of this message, the path DL data and serving eNB for the UE are changed. In the next step, the S-GW sends an acknowledgment message through the new path while the end marker sends this through an old path to stop the serving eNB from forwarding data. In the last three steps, the target eNB receives (Path-Switch-Req-Ack) from the MME as an acknowledgment, then the target eNB sends (UE-Context-Release) the message to the serving eNB via the X2 interface to release the resource.

2.5.7.3 S1-Based Handover Procedure

When the X2 interface does not exist between two network cells (different RAT), the S1- handover procedure will be used instead. Figure 2.15 illustrates the S1-based

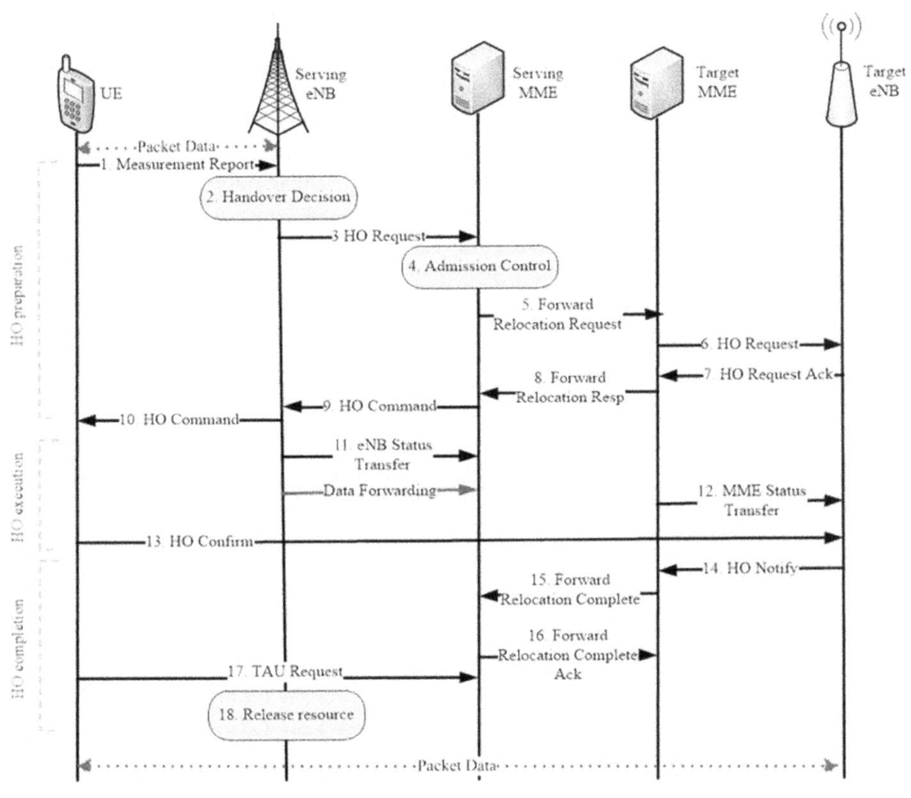

FIGURE 2.15 S1-based handover procedure. (Source: Sesia et al. 2011.)

handover procedure, which is more complex than the X2-based handover procedure, and is used to enable LTE\LTE-A to connect with both the wireless technology of 3GPP and the non-3GPP wireless technology (Miyim et al. 2014; Zhang et al. 2010). In addition, the S1-Based handover procedure has three phases, similar to the phases of the X2-based handover procedure (Sesia et al. 2011).

The preparation phase is included in steps 3.10, the execution phase is included in steps 11–13, and the completion phase is included in steps 14 to 18.

The measurement interval is the same as in the previous status. After handover decision is done, the preparation phase is created and the serving eNB sends (HO-Request) message to serving MME through S1 interface, then the serving MME performs admission control depending on the serving eNB message to increase the probability of handover success. Thereafter, the serving MME forwards the (Relocation-Request) message to target MME. In the next step, target MME sends (HO-Request) message to target cell through S1 interface. When the target cell receives the message and is accepted, then the target cell starts allocating the resources for new UE, and the target cell sends back the acknowledgment message (HO-Request-ACK) to the target MME through S1 interface. When the serving MME receives the (Relocation-Response) message from the target MME and accepts it, the serving MME sends (HO-Command) message to the serving eNB through the S1 interface. Then, the serving eNB instructs the UE to start execution of handover.

In execution phase, the serving eNB sends an (eNB-Status-Transfer) message to the serving MME and the serving MME sends (MME-Status-Transfer) message to the target cell through the target MME. The serving eNB starts forwarding the buffer packets to the target cell, and next when UE attaches to the target cell then the UE sends (HO-Confirmation) message to the target cell. Subsequently, the completion phase starts when the target cell sends (HO-Notify) message to the target MME. The target MME sends the (Forward-Relocation-Complete) message to the serving MME, and the serving MME sends (Forward-Relocation-Complete-ACK) message to the target MME to confirm acceptance. Then the UE sends Tracking Area Update (TAU-Request) message to the target MME for updating information and the serving MME sends the (Release-Resource) message to serving eNB to release all resources which are used for handover procedure.

2.6 EXISTING TECHNIQUES FOR MOBILITY MANAGEMENT IN LTE-ADVANCED WIRELESS SYSTEM

Due to the importance of a seamless handover for applications in LTE-A, new challenges have emerged with these improvements. Moreover, the LTE-A handover procedure is purely based on the HHO type, along with this line, the interruption handover time, data loss and buffering will be in UE plane. Therefore, the wrong selection of target cell during handover decision results in poor the QoS in the real-time application in term of high packet delay, high packet loss, and unnecessary handovers number. Thus, the reducing of the packet delay, the packet loss, and handovers number will affect the network performance by increasing the efficiency.

2.6.1 Horizontal Handover Decision Techniques in LTE\LTE-A

In LTE\LTE-A, the handover is initiated by the serving cell with assistance from the UE. The RSRP and RSQ measurements are done by the UE and are regularly sent to the serving cell. Moreover, the handover parameters perform the trigger of the handover procedure to optimize the goals of handover design. In addition, additional parameters are used to enhance handover decision and performance, such as the UE velocity parameter in (Lee et al. 2010; Yang et al. 2015), the traffic type parameter, which is based on the priority of data (real time or non-real time) in (Lee et al. 2010), the location parameter in (Chang et al. 2013; Chen et al. 2014b; Fei & Fan 2012), and the handover performance indicators in (Balan et al. 2011; Jansen et al. 2010; Yilmaz et al. 2013). In this subsection, horizontal handover decision algorithms are categorized into two groups based on algorithm mechanism and input parameters. These two groups are illustrated as mobility pattern history based and movement direction and location based.

2.6.1.1 Mobility Pattern-Based Algorithms

The mobility pattern-based technique or the informational history technique for handover prediction usually depends on the daily itinerary of the UE within a particular geographical area (Ge et al. 2009; Tao et al. 2016; Wang et al. 2014b). For example, the UE roams from his home to his work every morning within the same route. After finishing his work, he goes to a hot place (a hot place is any place except home and work). Every evening, he returns back home using regular routine routes. According to this information, the network gets the idea of the UE daily itinerary and then employs a handover optimal solution for future UE itinerary, and this causes the handover performance to be seamless and fast. This technique is beneficial in terms of avoiding a ping-pong handover, reducing too-early and too-late handovers, and reducing the cost of candidate cells evaluation; while the disadvantage of this technique is the cost of mobility database.

Tao et al. (2016) proposed an intelligent mobility pattern recognition, which is called the Smart HandOver (Smart-HO) technique, consisting of three phases: mobility pattern modeling, mobility pattern recognition, and intelligent handover implementation, respectively. This technique is employed in order to reduce ping-pong handovers, reduce handover time delays, and avoid redundant handover numbers. The authors started by creating the mobility pattern model through the tracking of UE movements inside network cells in the UE daily itinerary. Then, these movements were analyzed based on a time-domain model (at what time the UE accesses the network cell and for how long the UE has been staying in the network cell) to classify it within three movement cases: ping-pong, random, and linear. A 4-tuple table (current mobility pattern) will be constructed at the end of this step, which contains the sequence of network cells, the moment of access, the dwell time, and the probability parameters. In the mobility pattern recognition phase, the UE is observed to be roaming through neighbor network cells. This is done by comparing the UE path to the current mobility pattern to determine whether or not the UE is moving within the daily itinerary geographical area.

In the intelligent handover implementation phase, the author customized optimization strategies for two types of mobility patterns: the UE's ping-pong and linear movements, since the random mobility pattern could be ping-pong or linear. Depending on the ping-pong mobility pattern, the authors reduced the occurrence of ping-pong handovers by duplicating the dwell time of previous access cells in the table while moving between adjacent cells to avoid reconfiguring it when ping-ponging occurs. In the linear mobility pattern, the UE moved to a new network cell, which is already known, and thus the UE will start the pre-registration and pre-authentication process before starting the handover process to reduce latency.

Another standard technique for the mobility pattern was used in Ge et al. (2009) and Wang et al. (2014b). The authors used a UE mobility database under the law of UE mobility action for assisting fast and seamless eNBs to select target eNB, in order to minimize the handover number and ping-pong rate in LTE\LTE-A. Therefore, they used parameters that are related to cell_id, dwell time, the number of handovers, and handover probability of each UE pathway. However, the procedure starts when the UE is in active mode session and the measurement report is sent to the serving eNB. The UE compares the current pathway with the mobility pattern-based history to predict the target eNB, and then consequently updates the information of the mobility pattern-based history if needed.

On the other hand, Wang et al. (2014b) proposed some advanced parameters that are related to the time domain (day duration time, day connection time, connection frequency, and reconnection time), and that address the mobility database cost issue by storing its information in serving eNB and using the HO-Request message via an X2 interface to pass the information of the mobility pattern to other eNBs. In addition, the serving eNB is responsible for updating the mobility pattern history after the handover decision occurs, after which it passes the mobility pattern to the target eNB.

In contrast, the mobility pattern technique in Wanalertlak et al. (2011) depends on the short-term behavior of the grouping of UEs, where all UE movements are classified into many groups based on matching the time-domain with the location parameters. Therefore, the target cell is selected based on occurrence probability instead of RSS, and when the target cell selection fails, the next prediction will depend on another group.

2.6.1.2 Moving Direction and Location-Based Algorithms

In the case of the moving direction technique of handover prediction, the spatial information of eNBs and the UE location is important (Chang et al. 2013; Chen et al. 2014b; Hansen et al. 2009; Lee et al. 2005; Wang et al. 2014b). Therefore, Global Positioning System (GPS) is used to extract the position of the eNBs, the UEs, and the distance between the two. Depending on this type of information, which provides knowledge of actual network topology and the UE movement direction, the handover procedure performance can be enhanced, through reducing the number of handovers. On the other hand, the disadvantage of this technique is the unreliable accuracy of the GPS-extracted positions.

In Chang et al. (2013), Chen et al. (2014b), and Wang et al. (2014a), the authors proposed a handover decision technique based on tracking the last three positions of UE movement using GPS (the previous of previous (P_1), the previous (P_2), and

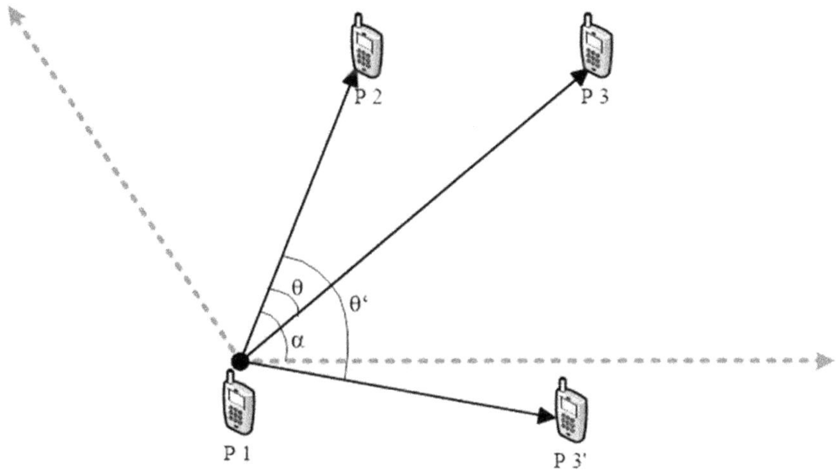

FIGURE 2.16 Moving direction prediction. (Source: Chang et al. 2013.)

the current (P_3)) to estimate UE future position with its candidate eNBs. The candidate eNBs are selected in the same UE direction by using a cosine function, and the target eNB is chosen from the candidate eNBs using a weight adjustment function. Moreover, this technique is used to reduce an unnecessary number of handovers. As shown in Figure 2.16, the UE current position is P_3, P_2 is the UE previous position, and P_1 is the previous of previous position.

The future position of UE will be in the same direction if θ is less than $|\pm\alpha|$, where θ is the angle of $\angle P_2, P_1, P_3$. Therefore, if θ is more than α, this means that the future movement will not be in the same direction as when θ' is between $\angle P_2, P_1$, and P_3'. In this case, P_2 becomes P_1, P_3 becomes P_2, and the UE will start looking for a new position for P_3. For the weight adjustment function in Wang et al. (2014a), the authors supposed a smaller θ with the nearest distance between UE and its candidate eNBs to select the target eNB, while in Chang et al. (2013) and Chen et al. (2014b), the authors supposed an incremental value for each candidate eNB weight for which the highest weight will be the target eNB.

Another technique for movement direction was proposed in Fei and Fan (2012). In this work, the authors suggested two schemes for an LTE-A with high mobility velocity. Scheme A was responsible for reducing handover delays through predicting the target eNB in the handover preparation phase even before the handover procedure starts. The authors supposed that the positions of neighbor eNBs and UE are known, and a new Prediction-Request message is sent to the neighbor eNBs depending on the distance value between the UE and the serving eNB for predicting the target eNB before a Measurement-Report message is sent. Then, the serving eNB selects the target eNB based on the shortest distance between UE, serving eNB, and neighbor eNBs. Therefore, the delay in handover preparation will be reduced by *dt*. Scheme B was responsible for minimizing unnecessary handovers through calculating the fading scale of each eNB surrounding the UE location. The authors chose the target eNB that belongs to the smallest fading scale value and the closest location; also, they added a positive offset parameter to the RSRP to avoid shadows and fast fading.

Another scheme that used the UE movement behavior for location prediction was proposed in Lei et al. (2011). The authors suggested the use of a spatial-temporal trajectory model, which consists of spatial information movements and temporal information movements in a suffix tree that shows the variation in probability between them. Therefore, this model follows a sequential regional relationship from the spatial scope and drives the UE time that matches the probabilities in the temporal scope. Based on this probabilistic suffix tree, the next movement of the UE can be captured. Their experiment proved that this model could improve prediction location accuracy.

According to the studies described in the previous subsection, which are also summarized in Table 2.2, all these algorithms aimed to provide efficient handover decision, but no optimal solution exists.

2.6.2 Vertical Handover Decision Techniques in LTE\LTE-A

In this subsection, the existing handover techniques in the two-tier access network are discussed. These five groups are illustrated as follows:

1. Power transmission loss based
2. Spatial information based
3. Mobility prediction based
4. RSS based
5. UE velocity based

This classification of handover decision algorithms in the two-tier networks is drawn through analyzing the main idea of the algorithm structure and the input parameters, which affect the outcome of the handover decision. The primary handover decision technique is used as a categorized criterion to simplify the comparison between the other techniques with regard to fundamental and operative understanding.

2.6.2.1 Power Transmission Loss-Based Algorithms

As previously mentioned, RSS is the primary parameter that triggers the conventional handover procedure. In the two-tier handover inside the E-UTRAN, the RSS indicator of eNB (high transmitting power) is always higher than the HeNB indicator (low transmitting power) when the location of HeNB is close to the eNB tower. Thus, Habibzadeh et al. (2015), Moon and Cho (2010), and Xu et al. (2013) proposed a modification to RSS algorithms to avoid the asymmetric power between eNB and HeNBs.

Moon and Cho (2010) proposed a new RSS-based algorithm for the hand-in handover. The idea of this algorithm is to discover the optimal combination factor by merging the RSS of the serving eNB and target HeNB for inferring a sensible handover criterion. Despite the good result in reducing asymmetry in the transmitting power, the performance of the system was decreased because of increased calculation complexity for each UE.

TABLE 2.2
Summary of horizontal handover decision reviewed algorithms

Scheme	Decision Input Parameters	Technique	Performance Matrices	Disadvantage
Mobility pattern-based				
Tao et al. (2016) Ge et al. (2009)	• DT, CT, CF • P • RSS Candidate cells list • Potential handover list	Information history of UE's mobility within intelligent recognition for determining the next UE roaming cell Information history of UE's mobility and simple mobility pattern matching are used in target cell prediction	• Reduce handover latency Improves throughput • Reduce ping-pong rate Reduce handovers number	• Increase the complexity of handover decision • The target cell selection is based on the probability
Wang et al. (2014b)	• RSRP, RSRQ • DT, CT, CF	Adding time-domain and connection frequency parameters to both UE's mobility information history and to the mobility pattern matching	• Reduce ping-pong rate Reduce handover failure rate	• Consumes both memory and power
Wanalertlak et al. (2011)	• UE-id • UE handover sequence • DT, CT	Grouping UEs' movements based on the location, connection time, and duration characteristics The target cell selection is based on the matching probability	• Reduce handover latency Improves prediction accuracy	• The complexity of handover decision is high
Movement direction-based				
Chang et al. (2013)	• P1, P2, P3 • RSRP • V, Θ, CLs • HFR, UHR	Selecting the target cell based on the UE moving direction, cosine function, weight adjustment function and UE's velocity	• Reduce handover failure rate • Improves handovers success rate	• Proposed solution does not consider cell capacity
Chen et al. (2014b)	• P1,P2, P3 • RSRP • Θ, D	Selecting the target cell based on the UE moving direction, cosine function, weight adjustment function and distance between UE and neighbor cells	• Reduce packet delay • Reduce packet loss • Improves throughput	• Does not consider the cell capacity

(continued)

TABLE 2.2 (Continued)
Summary of horizontal handover decision reviewed algorithms

Scheme	Decision Input Parameters	Technique	Performance Matrices	Disadvantage
Wang et al.(2014a)	• P1,P2,P3 • RSRP • D, Θ	Selecting the target cell based on the UE moving direction, cosine function, and weight adjustment function	• Improves packets loss • Reduce handovers number • Reduce packets delay	
Fei and Fan (2012)	• RSRP • D, Θ • CLs, P	The handover preparation starts before the handover procedure and the target cell is selected by geographical calculating	• Reduce packet delay • Improve handovers success rate • Reduce handovers frequencies	
Lei et al. (2011)	• P, V • DT, CT, CF	Predicting the future UE's location by using the variation of probability between spatial information movements and temporal information movements for UE in a suffix tree	• Reduce error prediction system • Improve number of patterns	The proposed solution is not practical in LTE-A

Note: DT: Duration time, CT: Connection time, CF: Connection frequency, RSS: Received signal strength, P: UE position, RSRP: Reference signal received power, V: UE velocity, Θ: the angle, CLs: Cell locations, HFR: Handover failure ratio, UHR: Unnecessary handover ratio, D: Distance.

Xu et al. (2013) solved this problem by considering the contradiction between the transmitting power of eNB and HeNB cells. The main idea of the algorithm is based on the incorporation of the RSS and path loss factors in both the serving eNB and target HeNB. The result showed a reduction in handover frequency and an improvement in HeNB utilization compared to the algorithm of Moon and Cho (2010). However, the traffic condition of HeNBs was not considered in these studies. Therefore, Habibzadeh et al. 2015) proposed a new handover decision algorithm based on traffic condition and propagation metrics. The main idea of this algorithm is the merging of RSS with the traffic factor (the expected service time) of the serving eNB and target HeNB cells. The handover is triggered in two ways. The first one is when the average RSS of the serving eNB is less than the threshold value for which the average RSS of the target HeNB is higher than that of the serving eNB plus hysteresis. The second one is based on service time. When the time duration estimation of the UE in the HeNB is longer than the channel holding time, the UE will connect to the HeNB. The result showed a reduction in unnecessary handovers and an increase in the probability of HeNB connection.

2.6.2.2 Spatial Information-Based Algorithms

Another different technique was introduced based on the spatial information of the UE and a possible adjustment of HeNBs, such as the neighbor HeNB location, the coverage area of each HeNB, and the distance between the UE and expected HeNBs. This technique can be obtained via GPS.

Liu et al. (2012) added the distance parameter to the handover decision scheme between HeNB and eNB. The proposal supposed a straight UE moving path in the circular geometry of the HeNB coverage, after which the distance between the UE position and neighbor HeNBs location is calculated. The calculation equation is based only on the RSS change rate in different UE positions to estimate the UE velocity. The authors also defined the distance threshold value to indicate whether or not the UE should perform the handover. When the UE distance value is less than the moving distance, the UE will launch the handover decision scheme. Based on this technique, unnecessary handovers were reduced through the use of numerical simulation.

Sung et al. (2013) proposed a new handover scheme based on the awareness location of macro cells, HeNBs, and UE position to reduce candidate HeNB numbers. The proposal traced the UE movement to predict the UE future location. Then, parameters of time were used to group together the UE with the movement space. In this way, the duration of time in which the UE stays connected to the target HeNB can be estimated. If the excepted UE connection time is less than the value of predefined time to stay for the neighbor cell, the neighbor cell will ignore this parameter from the UE candidate list. The aforementioned procedure is done by the UE serving cell. After the candidate list is created, the UE will select the target cell based on the analysis of power transmission, interference, and throughput of each cell in the candidate list. In summary, the proposed algorithm improved the handover procedure by reducing the handover number, but the complexity of the system still remained high.

Bilen et al. (2016) proposed a new handover management for vertical networks to reduce ping-ponging and unnecessary handover numbers. Their approach is based on the selection of an optimal eNB for the handover in which the eNB operates the control channels and the HeNB operates the data channels. The algorithm uses UE location and eNB quality indicators, which are collected from both eNB and HeNB to select the optimal eNB and HeNB for the UEs. The eNB indicators are used to draw the candidate cells. This is done through selecting the lowest gain first. Then, the Kriging Interpolator is used together with the Semivariogram Analysis to estimate future UE positions. Thirdly, the autoregressive model monitors each candidate cell in a certain timescale. These three operations are performed by the eNB node before triggering the handover. Hence, the optimal eNB is selected based on minimum variation.

Chang et al. (2012) proposed a new efficient signaling measurement algorithm when the UE moves from the eNB to the HeNB (hand-in handover). The proposed scheme is based on two functions: location-based neighbor eNB management and HeNB-assisted handover. The location-based neighbor eNB algorithm is used to reduce neighbor cells by eliminating the HeNB that is located outside of the UE trajectory range. The algorithm uses RSRP and RSQ to estimate the distance between the UE and its serving eNB, which includes the eNB sector-id. The HeNBs located at this intersection will be included in the candidate list for the handover. The HeNB-assisted handover function is used to assist the UE in determining the time of the handover. However, in both parts of the algorithm, the eNB and HeNB cells are required to preserve the UE trajectory or observe the uplink signal from the UE via the extra wireless interfaces of the eNB and HeNB cells.

2.6.2.3 Mobility Prediction-Based Algorithms

In the case of mobility prediction-based algorithms, the GPS is used to extract the eNB position, UE position, and the distance between them.

Ulvan et al. (2011) used mobility prediction to predict the future position of UE and to estimate the target HeNB to reduce frequent and unnecessary handovers. Their handover scenarios cover the hand-in, hand-between, and hand-out procedures. The authors used the Markov chain method for probability transition among cells and assumed two handover strategies, which are the proactive and reactive handovers. In the proactive handover, the handover is executed at any time without the need to verify the serving RSS. The network estimates the UE position before the UE arrives and the UE estimates the remaining time before the standard handover is triggered, after which the UE will trigger the handover early. On the contrary, in the reactive handover, the handover execution is postponed as long as possible. This strategy aims to minimize unnecessary handovers even when the UE has found a stronger RSS than its serving cell. The handover is triggered when the RSS value of its serving cell is lesser than the threshold value. In addition, the authors limited the handover to the HeNB according to the velocity and priority of data (real-time or non-real-time). However, this technique wastes memory and results in a high probability error of the target cell.

Maheswari et al. (2016) used mobility prediction to predict the future position of UE and to estimate the target HeNB to reduce handover latency. GPS was used to track the UE positions and update system discovery. Moreover, the Markov chain function was also used to predict the best target HeNB for the handover. In addition, on the research of mobility prediction, Jeong et al. (2011) based their proposal on the mobility history and location prediction of the handover decision to eliminate unnecessary handover numbers. The algorithm estimated the UE time of stay in HeNB so as to avoid temporary HeNB visitors. The authors used previous UE movement patterns and predefined movement sequences to predict the time interval of future HeNBs.

2.6.2.4 RSS-Based Algorithms

In the standard handover procedure, the scan of RSS measurement is done periodically for all neighbor cells. Therefore, the UE power consumption is based on RSS measurement and the handover decision is based on the strongest RSS value even when the UE moving direction is opposite to the cell location. Accordingly, in Lin et al. (2016) and Marshoud et al. (2016), the authors modified the RSS measurement to improve the handover decision procedure.

Venkata et al. (2015) proposed a handover management approach to assist the handover procedure among HeNBs, which also reduced the handover number and led to void HeNB overloading. The authors assumed a table, which is stored in the UE, which has all the HeNB_ids and updated these in each RSS measurement scan. Thus, based on the RSS value of each HeNB, the UE serving cell will recognize the direction of the UE movement. Based on this table, the target cell will be the HeNB, which has the highest RSS, for which the antecedent RSS values will increase gradually. The handover is triggered after investigating if the serving cell RSS is lesser than the threshold and the RSS of the target HeNB is higher than the threshold, as well as ensuring that the target HeNB capacity is not full. However, the proposed algorithm consumes both memory and power.

Deswal and Singhrova (2016) improved the use of the RSS parameter in the standard handover through assuming different RSS threshold values for the HeNB. They increased the RSS threshold value for the handover procedure from HeNB to eNB. On the contrary, they decreased the RSS threshold value for the handover procedure from eNB to HeNB to prevent call drops by making the handover to eNB a little earlier.

2.6.2.5 UE Velocity-Based Algorithms

In velocity-based handover, a few researchers added UE velocity as an input parameter to the handover procedure to reduce the number of unnecessary handovers in vertical networks.

Wu et al. (2013) added a periodic trigger along with UE velocity to the standard handover procedure, so as to prompt the network system throughput. The periodic scan for triggering the handover together with UE velocity forces the UE to execute the handover within HeNB even if the RSS of the serving eNB is better than the HeNB RSS. Moreover, UE velocity is also considered in the handover decision to prevent handoff to HeNBs when the UE residential time is very short. Therefore, when the UE serving cell is the eNB, the RSS and periodic trigger scans will be

TABLE 2.3
Summary of vertical handover decision reviewed algorithms

Scheme	Decision Input Parameters	Technique	Performance Matrices	Disadvantage
		Power transmission loss-based		
Moon and Cho (2010) scheme.	• RSS	Merging the RSS of serving eNB with target HeNB to infer the sensible handover criterion.	• Reduce handovers number Improve the probability of handover	• Increases complexity of handover decision-making • Cannot be fully implemented
Xu et al. (2013) scheme.	• RSS	Incorporation of RSS and path loss factors in serving eNB and target HeNB	• Reduce handovers number	• Not implemented in a real environment
Habibzadeh et al. (2015) scheme.	• RSS	Merging RSS with traffic factor (expected service time) of serving eNB and target HeNB cells.	• Reduce handovers number	• Assumes constant UE speed in real environments
		Spatial information-based		
Liu et al. (2012) scheme	• RSS • Distance • UE speed	Selection of the target cell in handover decision is based on the HeNB distance prediction of UE movement	• Handovers failure • Reduce handovers number	• •
Sung et al. (2013) scheme	• RSS • UE positions • eNBs location • Time to stay	Selection of the target cell is based on the future UE position, time to stay, and the UE neighbor RSS cells.	• Reduce handovers number • Improve throughput	• Increases complexity

Moon and Cho (2010) scheme	• Call drop rate • Call block rate • Packet loss rate • Ping-pong • Unnecessary handover • UE location • Timescale prediction	An optimal cell selection mechanism for handovers is based on spatio-temporal estimations	• Reduce ping-pong rate • Reduce handovers number	• Searching time during target cell selection through calculating many input parameters
Chang et al. (2012)	• RSS • UE and cells location	A new transmission measurement is used in handover decision procedure based on the UE moving direction and cells' location	• Reduce transmission measurement rate Reduce handovers number	
		Mobility prediction-based		
Ulvan et al. (2011)	• RSS • UE and cells location • UE speed	Select the target cell in handover decision based on the Markov chain method for UE's probability transition among cells and within two handover strategies which are proactive and reactive.	• Reduce handovers number • Reduce handover latency	• The cost of mobility database is much high through using destiny of HeNBs
Maheswari et al. (2016)	• RSS • UE and cells location •	Select the target cell in handover decision based on the Markov chain method for UE's probability transition among cells	• Reduce packet loss • Reduce handover delay	
Jeong et al. (2011)	• RSS • UE and cells location • Time to stay	The mobility history and UE's Future location prediction are used in handover decision procedure to select a target cell	• Reduce unnecessary handover • Reduce error probability	• Not a comprehensive solution

(continued)

TABLE 2.3 (Continued)
Summary of vertical handover decision reviewed algorithms

Scheme	Decision Input Parameters	Technique	Performance Matrices	Disadvantage
		RSS-based		
Venkata Rajesh M et al. (2015)	• RSS	The UE's serving cell knows the direction of UE movement based on the gradation of RSS value for each UE's neighbor cell then assists handover decision procedure in selecting a target cell	----	• Consumes both memory and power • Increase interruption time
Deswal and Singhrova (2016)	• RSS	Used various RSS threshold values in handover between eNB and HeNBs	• Reduce handovers number • Reduce packet loss	• Increase interference
		UE speed-based		
Wu et al. (2013)	• RSS • UE speed	Added speed parameter to handover decision procedure for selecting target cell	• Reduce unnecessary handover	• Increase interference

considered in handover triggering. On the other hand, only RSS is considered in handover triggering when the UE serving cell is the HeNB.

According to the studies mentioned in the previous subsection, also summarized in Table 2.3, all the proposed algorithms aim to provide an efficient handover decision in two-tier networks, but as yet no optimal solution exists.

2.7 SUMMARY

HHO is a handover technique used in LTE-A systems to enhance the system throughput at high-velocity mobilities through reducing the handoff complexity. In the HHO technique, handover decisions are distributed between the UE and serving cell without the intervention of core network. Thus, the probability of data loss and delay is high because of handover control distribution. Therefore, the mitigation of handovers number, packet loss, and packet delay will improve the LTE-A system throughput in both types of networks (homogeneous and heterogeneous). In this chapter, the types of networks in LTE-A and the handover procedure were explored. Moreover, the handover challenges were discussed and the handover decision algorithms were surveyed. In addition, a taxonomy of handover decision was proposed in both types of networks based on input parameters and techniques. Consequently, summary tables were created to illustrate the existing works with regard to the handover decision in a simple way.

The classification of handover decision algorithms in the current chapter will be very helpful to the research community and serve as a considerable introductory material for someone who embarks on handover techniques. Based on this survey, various existing handover decision solutions have been introduced to cover certain phases in the handover procedure in order to reinforce QoS for which the results of these studies can be improved. Moreover, the standard HHO is still incapable of dealing with networks in high UE velocities and high HeNBs density. In order to overcome such unique handover decision challenges, the consideration of new input parameters is compulsory in designing HHO procedure. In addition, these solutions entirely or in part are combined to build an effective overall solution. For instance, merging between information history and moving direction techniques of the UE for improving the handover process will enhance the QoS by decreasing the unnecessary handovers and handover latency.

3 Enhancements in Techniques for Mobility Management in LTE-Advanced Wireless System

3.1 INTRODUCTION

The handover technique in both types of LTE-A networks (the horizontal and vertical) is a field of research, which still needs improved and enhanced system performance. This chapter presents the potential approach for enhancements to overcome the constraints in the subject domain, as mentioned in Chapter 1. The approach for improved techniques for Mobility Management in LTE-Advanced Wireless System begins with an investigation into the possible gaps in the handover decision procedure. This is followed with determining the weaknesses in the handover decision procedure. Moreover, this chapter investigates additional parameters that enhance the handover decision procedure and performance. The approach for enhancement is expounded and discussed in general in Section 3.2. The LTE\LTE-A handover procedure is explained in Section 3.3. The specifics of the proposed solutions and algorithms are discussed in Section 3.4. The simulation model to validate the proposed solutions and algorithms is presented in Section 3.5. The performance evaluation for the proposed solutions and existing works are discussed in detail in Section 3.6. Moreover, the input parameters threshold values are validated to select the best threshold values for both algorithms in Section 3.7. Finally, the chapter summary is clarified, and the approach adopted for enhancements in subject of this book concluded.

3.2 GENERIC APPROACH FOR ENHANCEMENTS

The generic design of approach for enhancements in this research domain start with the problem formulation. This includes the identification of the fundamental problems faced and parameters used to assess system performance. Problem identification is also meant to justify investigation into the related work. It also allows for in-depth dissection of the research scenario concerning the coordination between the handover procedure and LTE-A infrastructure network. Next, problem formulation is accompanied with the proposed solutions that would improve total system performance. The validation of the developed methods and solutions is performed through developing a simulation model that resembles real-time communication scenarios. The simulation model is the backbone of the validation and evaluation process for

DOI: 10.1201/9781003648130-3

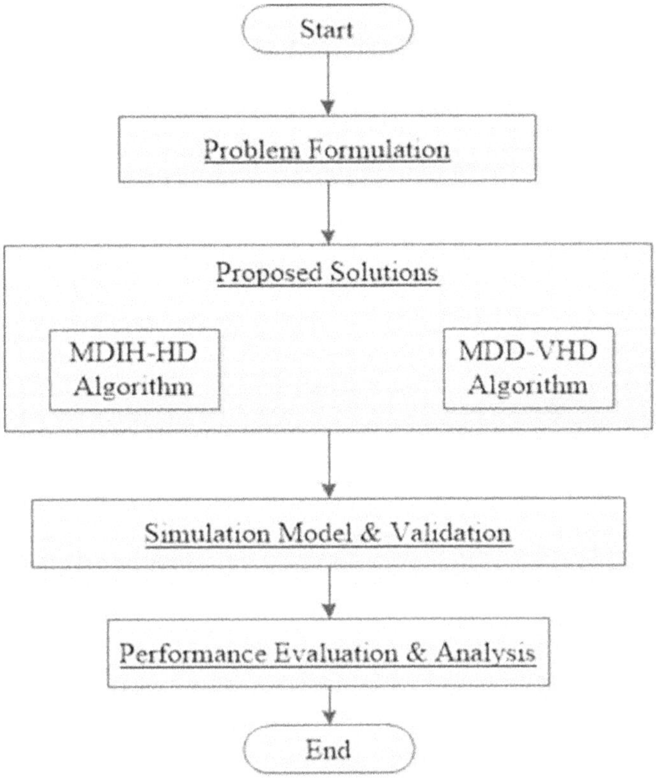

FIGURE 3.1 General approach for enhancements.

the proposed methods. Next, benchmarking the proposed handover decision algorithm against the standard and existing works with regard to different metrics of performance is also done. This structure of the approach for enhancements in existing techniques for mobility management in LTE-advanced wireless system is further clarified in Figure 3.1.

3.3 THE LTE\LTE-A HANDOVER PROCEDURE

The procedure of the 3GPP handover decision algorithm is explained in Figure 3.2.

This handover decision algorithm is part of the HHO procedure, which is adopted in the LTE-A system (3GPP 2012b; 3GPP 2016a) and described in Chapter 2. Initially, the UE establishes a connection with the appropriate serving cell and the carrier that provides the strongest signal. After connection establishment, the signal quality as perceived by the UE will be monitored for use in the handover decision (Han & Wu 2010). This process is performed periodically within a preconfigured time interval $T_{measurement}$. Upon the estimation of signal quality, the UE sends a measurement report to the respective serving cell. The serving cell uses the measurement report to select the best cell as a target cell. If the handover decision criteria are fulfilled based on the

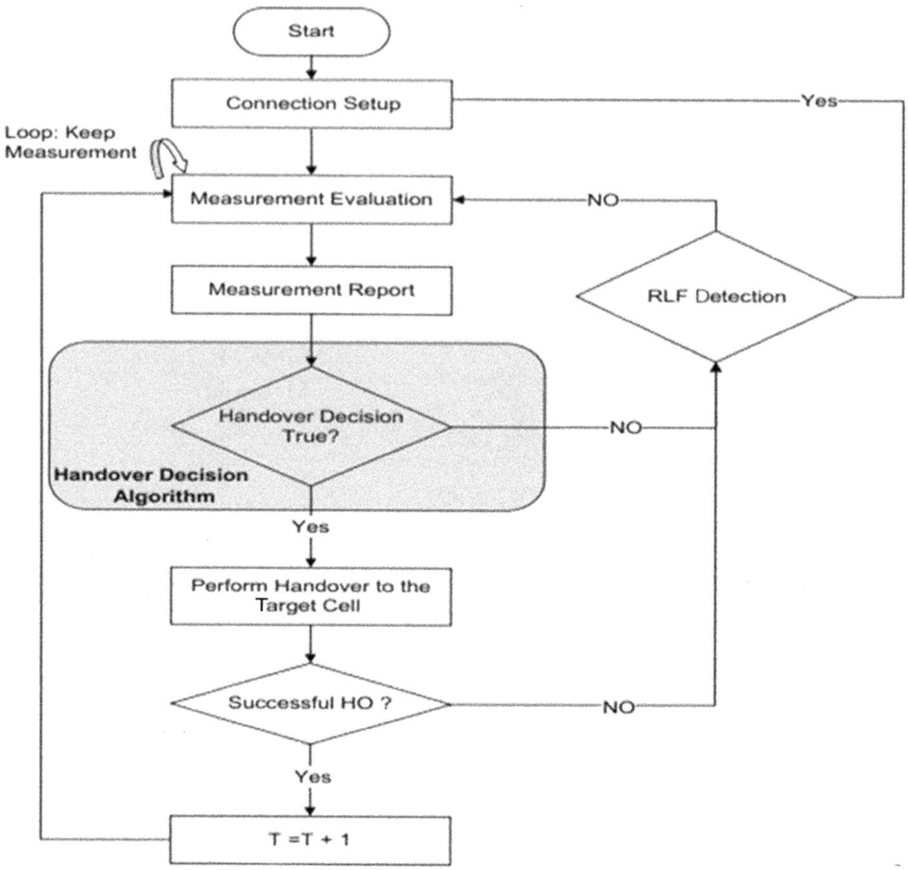

FIGURE 3.2 The 3GPP handover decision algorithm flowchart. (Source: 3GPP 2016a.)

measurement report and handover control parameters, the serving cell will take over the handover decision. The handover procedure will be commenced upon the verification of the handover decision criteria by the serving cell. Upon successful handover, the UE will continue with the next cycle of parameter estimation. Conversely, if the handover decision is not met, the UE will maintain the connection with the serving cell as long as the signal quality is satisfied. Otherwise, the RLF detection will be triggered.

3.4 PROPOSED SOLUTIONS

The proposed solutions for enhancements are considered as the heart of this work and are suggested to support the handover performance in two types of LTE-A networks (horizontal and vertical). In this work, user profile is used to assist the handover parameters of the handover decision to enhance the LTE-A system performance. In general, the proposed architecture of the LTE-A system and the user profile parameters in the handover decision are further described in Figure 3.3.

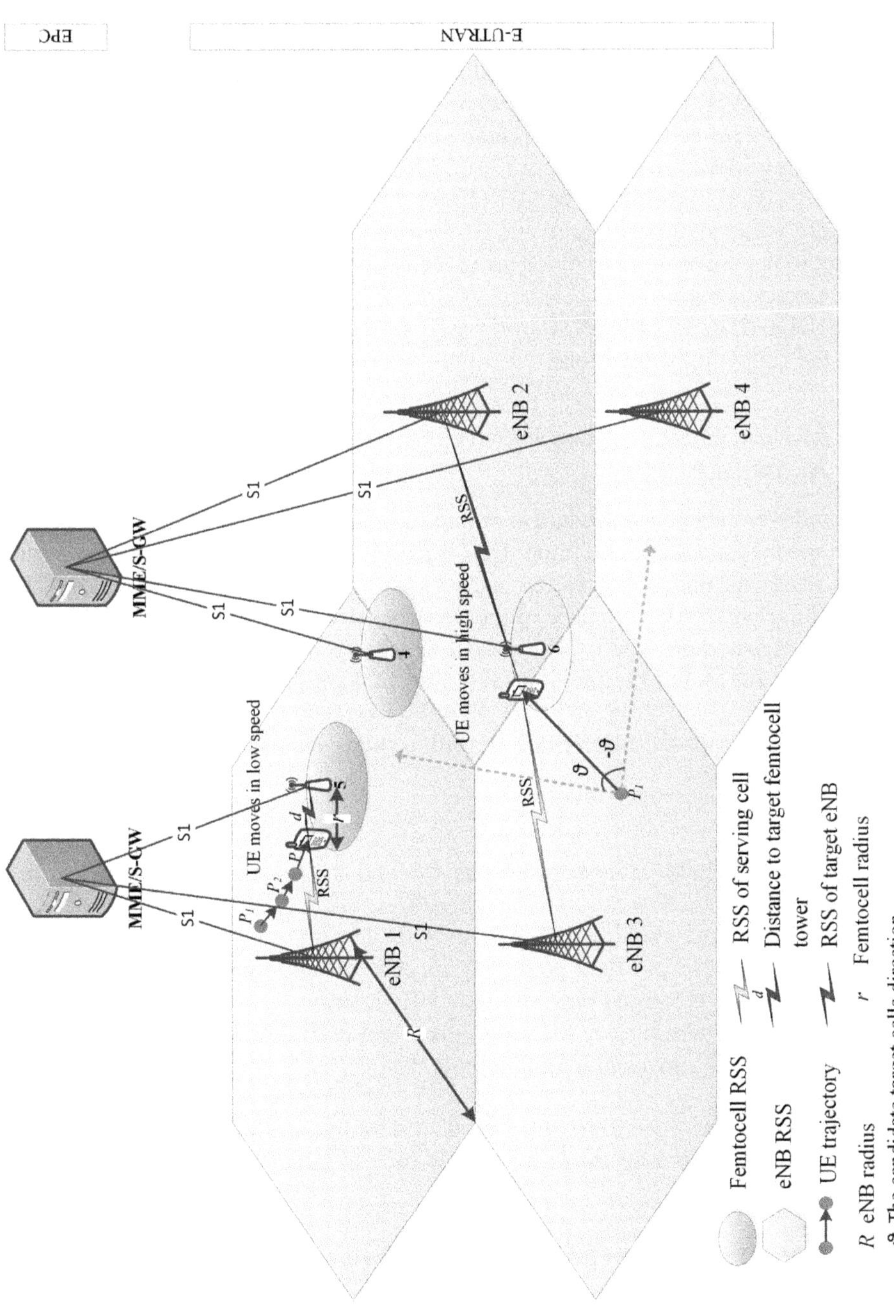

FIGURE 3.3 LTE-A system architecture and user profile parameters in the handover procedure.

The handover decision is assisted via the user profile, such as input parameters. This is to select the target cell in a short time. The UE direction assists the handover decision through avoiding signaling measurement controls with neighbor cells that are not located ahead of UE trajectory as well as in selecting the neighbor cell that fits as the target cell. Thus, the two proposed algorithms in the handover scenario (horizontal and vertical) are based on GPS technology. Each UE position and neighbor cell location are determined using GPS (Huang et al. 2013). Moreover, the UE velocity (*V*) is also considered in the selection to determine the suitability of the proposed handover decision algorithms (MDIH-HD or MDD-VHD) for handover decision procedures (Wu et al. 2013). The *V* and RSS parameters aid the handover decision in reducing the handover number by clearing HeNBs from the candidate cell list when the UE velocity is high.

The MDIH-HD and MDD-VHD proposed structure algorithms are explained in detail in the following subsections.

3.4.1 Moving Direction Informative Historical Handover Decision Algorithm

A new handover decision procedure is proposed in this work by taking into consideration UE velocity, prediction of future UE direction, and UE routine routes along with standard handover parameters (reference signal received power [RSRP] and TTT). This is to improve the performance of the handover decision between the eNBs when the UE velocity is high. The UE knows its velocity from using the GPS technology (Huang et al. 2013) and the previous and current positions of UE are saved in the UE cache.

In detail, the proposed MDIH-HD algorithm improves handovers by considering the following criteria:

1. UE velocity.
2. RSRP threshold value between the UE and its serving eNB to execute the handover.
3. UE direction: tracking UE positions using two points (the previous and current points) and then predicting its direction in order to select the target eNB, which is located in front of the UE direction.
4. Mobility pattern history to predict the target eNB directly from the UE table when the UE moves in a routine route.

Moreover, the cosine function is also used in determining the candidate eNB list. Meanwhile, the weight adjustment function is used to select the target eNB from the candidate eNB list.

3.4.1.1 Moving Direction Prediction Procedure

As the UE moves ahead of neighbor eNBs, and when the RSRP among the UE and the serving eNB is less than the threshold, the previous and current UE positions are

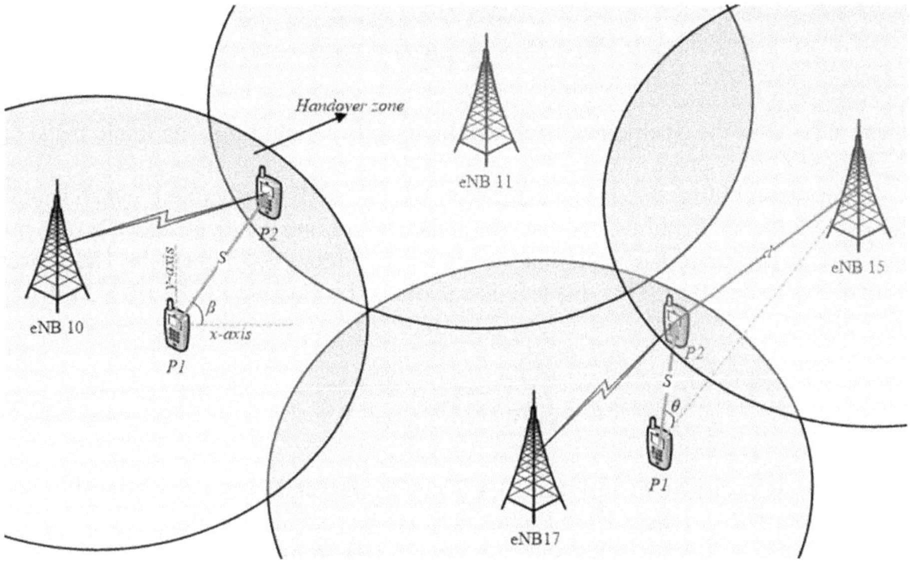

FIGURE 3.4 The UE moving direction and candidate eNBs selection.

reviewed from the GPS and represented by P_1, P_2 to determine future UE directions. Figure 3.4 illustrates the P_1 and P_2 positions when the UE moves inside eNB 10.

Based on Figure 3.4, the $P_2(x_2, y_2)$ position is selected close to the border of the serving eNB (eNB 10) where the RSRP between the UE and serving eNB is less than the threshold. $P_1(x_1, y_1)$ is tracked frequently based on distance S, where distance S should be equal to a threshold value. β is the angle of UE direction, which is utilized to estimate the UE future direction by analyzing the slope between the previous and current UE positions, as gained from Equation (3.1).

$$\beta = \tan^{-1}\left(\frac{y_2 - y_1}{x_2 - x_1}\right) \tag{3.1}$$

In addition, in order to avoid power consumption and squandering time when UE sends a measurement report to its serving eNB, two filters were applied for choosing the candidate eNBs. In the first filter, the nominee eNBs should be located ahead of the UE direction within the range of $| \pm 60°|$, and each eNB that is situated inside this range will have the chance to be incorporated into the candidate eNB list (Wang et al. 2014a). This value range was chosen based on the fact that each eNB has three array antennas, which cover 360° of signal coverage. Moreover, each array antenna supplies 120° of signal coverage. Consequently, every neighbor eNB of the UE was verified through calculating the angle of $\angle P_2, P_1, neighbor{-}eNB$ to be in that range; otherwise, it will be disregarded. This angle is calculated by Equation (3.2) (Almulla et al. 2014; Lu et al. 2014).

$$\theta = \cos^{-1}\frac{(C-P_1).(P_2-P_1)}{|C-P_1||P_2-P_1|} \tag{3.2}$$

where $C(x, y)$ is the position of each neighbor eNB cell and θ is the angle between P_2, P_1, and C.

In the second filter, the distance between UE and eNB was applied, which should not exceed the eNB radius, so that the eNBs, which are far from the UE, are not included in the candidate eNB list, as shown in Figure 3.4, when the UE moves from eNB17 to eNB15. This distance is calculated by Equation (3.3) (Ben-Mubarak et al. 2013).

$$d_c = \sqrt{(x_3 - x_2)^2 + (y_3 - y_2)^2} \tag{3.3}$$

The verification of eNB15 in Figure 3.4 depends on θ and distance values; if $\theta \leq |\pm 60^{\circ}|$ and $d \leq eNB_{radius}$, the eNB15 is considered to be a candidate eNB.

The next step involves choosing the target eNB from the nominee eNB list by using the Weight Adjustment algorithm (Wang et al. 2014a). In the proposed model, the closest distance to the UE's present position and the smallest θ from the candidate eNB list will be the most suitable for use as the target eNB. The Weight Adjustment algorithm is shown in Algorithm 3.1.

Algorithm 3.1 Weight Adjustment

1: Input θc and dc

2: Output: W

3: Anorm= 1- $\frac{\theta c}{\alpha}$

4: Dnorm=1- $\frac{dc}{2r}$

5: W= Anorm + Dnorm

where θc is the angle of eNB, dc is the distance between eNB and UE's present position (P_2), and W is the summation value, which is considered for selecting the target eNB

Moreover, normalization was also performed for both the angle and the distance, so that both would be according to standard unification. α is the angle value that is used for normalization. *Anorm* contains the outcome of the angle normalization and because all the candidate eNBs' angles are less than or equal to $|\pm 60^{\circ}|$, this angle (α) was used for the normalization process. *Dnorm* contains the outcome of distance normalization, and this was normalized using the eNB transmission range ($2r$) to increase the angle priority, as the distance of all nominee eNB list is less than or equal to r.

3.4.1.2 Historical Connection Patterns

According to the moving direction prediction procedure, the signaling measurement consumption and checking time should be sharply mitigated through minimizing the nominee neighbor eNB list. In this section, the historical UE itinerary connection pattern method is employed to select the target eNB directly without going into the measurement stage.

In this work, the information of the UE trajectory for each ongoing handover was recorded through the moving direction prediction procedure one at a time. Table 3.1 shows the history of the UE itinerary connection patterns called *Tb*, which comprises record number (*RecNo*), serving eNB (eNB_S), target eNB (eNB_T), UE slope angle (β), and the UE present position ($P_2(x_2, y_2)$). Furthermore, these attributes will act as a reference for upcoming UE movements in an urban scenario, where most of the UEs are used in the same itineraries every day.

In Figure 3.5, by accepting that the UE database has not been filled, the UE moves from its home to work for the first time. At the point when the UE reaches the edge of its serving eNB (*eNB*1), the moving direction prediction procedure will be called. The first record in the table of Figure 3.5 will be made depending on the previous procedure after which the handover decision will be triggered on eNB9. In eNB9, a similar method will be applied when the UE moves to eNB7 and so the second record will be added to the table. Therefore, in the upcoming time when the UE uses the same path with the same serving eNB, the target eNB will be selected from the UE table. The target eNB is selected based on matching the present UE path (eNB_S, β, P_2) with what is included in its table (*Tb*) in order to observe whether or not there is a current record; and afterward returning back the target eNB. On the other hand, if the current UE trajectory does not match any record in its table, the UE will apply the moving direction prediction procedure and add a new record. The UE table acts like a small map for each handover process, and each UE starts by matching its moving direction with what is included in its table when it arrives at the P_2 position.

3.4.1.3 Moving Direction with Historical Connection Patterns Algorithm

Both techniques of the moving direction prediction procedure and historical connection patterns for MDIH-HD are represented by Figure 3.6. The execution of the MDIH-HD algorithm starts when the UE moves inside its serving eNB but does

TABLE 3.1
The UE historical connection patterns

Rec No	eNB_S	eNB_T	β	$P_2(x_2, y_2)$
1	eNB1	eNB9	20°	2,999,101.7

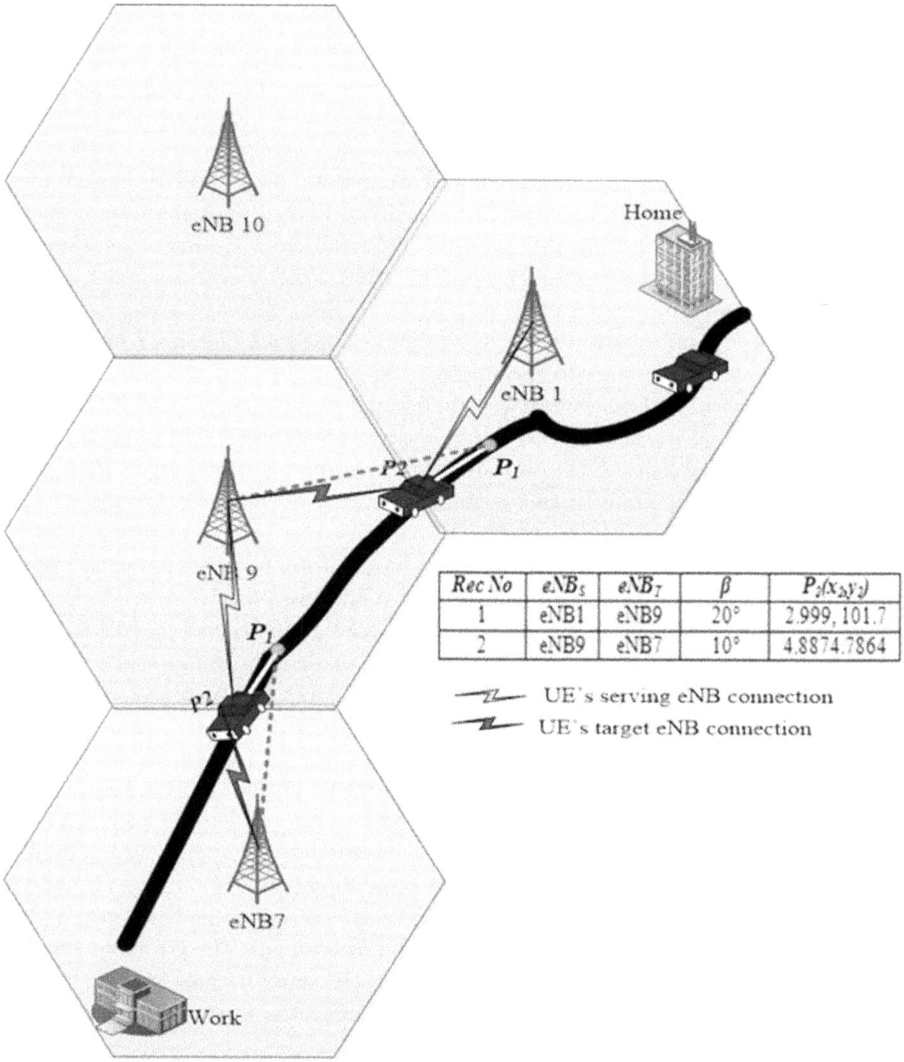

Rec No	eNB_S	eNB_T	β	$P_2(x_2, y_2)$
1	eNB1	eNB9	20°	2.999, 101.7
2	eNB9	eNB7	10°	4.8874.7864

FIGURE 3.5 Historical UE's path connection table.

not come close to the serving eNB border; the P1 position is picked up every distance *S* and stored in the UE cache. *S* is the distance threshold value that is used to keep appropriate spaces between the traced UE positions.

When the UE is close to the border of its serving eNB and the RSRP in between is less than the threshold, the P_2 position is reported by the UE. The UE starts calculating the slope (β). After finding β, the UE starts searching in *Tb* to find a record, which matches with its current moving direction (*serving eNB – id*, β, $P_2(x_2, y_2)$), where *Tb* is a table which represents the historical UE trajectory connection pattern database and is stored in the UE memory. The searching starts by matching the current

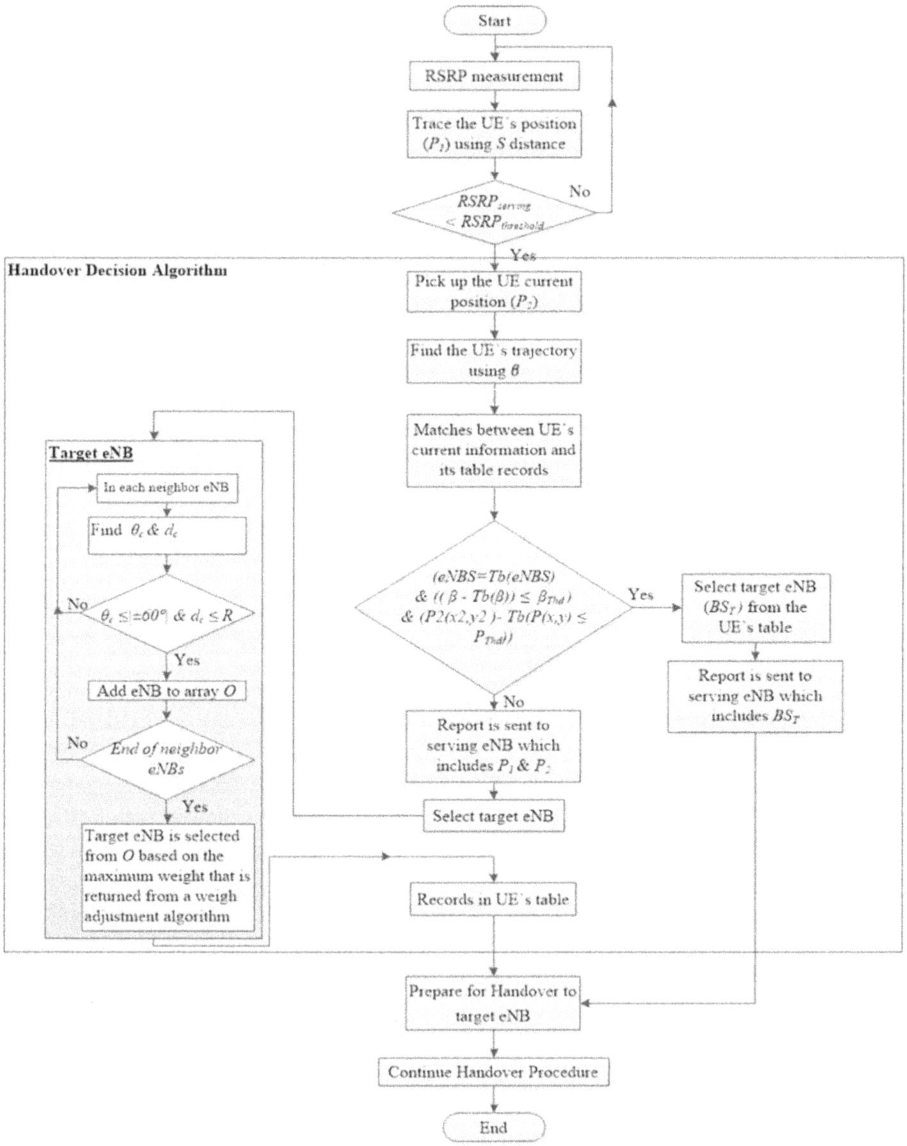

FIGURE 3.6 The MDIH-HD algorithm.

servingeNB – *id* with the same id in *Tb*, and then the UE matches the current β with the nearest β in *Tb* by using a threshold of this approximation; also, for the current P_2 position, an approximation is applied to match the values of the current P_2 and $Tb(P)$ in order to give the UE a little free space away from its previous trajectory. If there is a record that matches with the current UE moving direction, it will return back the target eNB to the BS_T variable; otherwise, the BS_T variable will be empty. In the next step, the UE sends a report message to its serving eNB, which includes the P_1, P_2, BS_T parameters.

Meanwhile, if the eNB_T is empty, the serving eNB starts searching for the target eNB. The first step is to select the candidate eNBs from all neighbor eNBs by calculating the angle and distance of each one. The calculation is based on the UE direction (P_1, P_2) in addition to eNBs, which are located ahead of UE and suit the$|\pm 60°|$ range, in which the distance should be less than or equal to (r). The result is a O array, which includes all nominee eNBs. The next step is to select the target eNB from O based on the smallest angle and the shortest distance. The distance between UE (P_2) and each eNB in O is calculated and stored in variable d_c, and then sent along with the eNB angle to the Weight Adjustment function to calculate the weight. Each candidate eNB has weighted, and BS_T is the eNB that has the maximum weight. After selecting the BS_T, the serving eNB sends a report to UE, which contains BS_T and so adding a new record to its table. The RSS-procedure is triggered when the RSRP value of the serving eNB is less than the target eNB (BS_T) RSRP value within the TTT period; the handover occurs in BS_T; after the handover is completed, the target eNB will be the serving eNB for UE, the P_2 position will be P_1, and both P_2 and BS_T will be empty.

3.4.2 Moving Direction and Distance in Vertical Handover Decision Algorithm

A new handover decision procedure is proposed related to the vertical networks based on the distance between the current UE position and HeNBs location. As the transmission power of eNB is allowed to be higher than the transmission power of HeNB, the standard handover which is based only on RSS parameter is not efficient. Therefore, new input parameters are also taken into consideration together with the standard handover parameters (RSRP and TTT).

In addition, MDD-VHD suggests a new scheme which improves the handover performance in the three vertical handover scenarios (hand-in, hand-out, and inter-HeNBs). The proposed algorithm is based on these main factors:

1. UE velocity: to support QoS for low-velocity users.
2. RSRP threshold value between UE and its serving eNB.
3. Tracking UE positions: to predict the future UE direction through using the polynomial function for the previous different UE positions to clear out-range neighbor cells.
4. The distance between current UE position and HeNBs location: to force the UE to handoff to the front HeNBs if the distance is less than the predefined value.

Moreover, the cosine function will be used for candidate cell list while the weight adjustment function will be used for selecting the target cell from the candidate cell list.

3.4.2.1 The UE Velocity

This work considers the maximum UE velocity which is needed to handoff to HeNBs to be 30 km/h (Wu et al. 2013; Zhang et al. 2010).

3.4.2.2 The Distance between UE Position and HeNBs Location

In this work, the HeNB coverage area is considered to be a circular geometry cell and the eNB coverage area to be a regular hexagonal cell. Moreover, the deployment of HeNBs will fall under the eNB coverage area, for which the coverage of HeNBs will be quite small compared to the eNB coverage area. Furthermore, the variation in the transmitting power between both of them would be quite high, especially for the HeNBs, which are close to the eNB tower. Thus, add the periodically scanned distance (d_c) along with the RSS in the UE measurement control. As the UE moves toward the HeNBs and while the UE measurement control finds the distance (Equation 3.3) between the UE and HeNB location to be less than the predefined value, the measurement report will be sent to the serving cell to trigger the handover procedure, as shown in Figure 3.7.

Based on Figure 3.7, the UE moves from $P_1(x_1, y_1)$ to $P_4(x_4, y_4)$ and when reaching P_4, the UE measurement control finds the distance (d) between the UE position and *HeNB*1 location (or must be less than the threshold value (r)) where r is the radius of the HeNB. The UE will then hand off to HeNB1.

3.4.2.3 The UE Trajectory Prediction and Candidate Cells Selection

In order to predict the future trajectory of the UE through using the previous UE positions, the polynomial function will be used for each UE to describe its own pathway (Chatterjee et al. 2012). Based on Figure 3.7, P_1, P_2, P_3, and P_4 are considered to estimate the future UE direction. The Queue (Q) is used to trace the newest four positions of UE, and the future direction of the UE is based on the polynomial equation for these four positions. The polynomial function $f(x)$ generates four positions and each position is saved in the UE's cache as (x_i, y_i), where i is from 1 to 4. The polynomial function in Equation (3.4) is used:

$$f(x) = a_o x^j + a_1 x^{(j-1)} + a_2 x^{(j-2)} + \ldots + a_j \tag{3.4}$$

where j is the last UE positions and $a_0 \neq 0$. In order to predict the UE direction when the handover starts launching, a position (Ps) is supposed. Accordingly, the $Ps(x)$ is between $P_1(x)$ and $P_2(x)$, while the $Ps(y)$ value is the output of the polynomial function. Subsequently, analysis of the UE trajectory will be based on the linear prediction between $Ps(x, y)$ and the current UE position $P_4(x_4, y_4)$, as illustrated in Figure 3.8.

Moreover, when the UE sends the measurement report to its serving cell, the measurement report message contains the UE trajectory information that includes the Ps and $Q(last)$ coordination, where $Q(last)$ is the current UE position. In addition, to avoid time wastage during hand-in, inter-HeNB, and hand-out handovers,

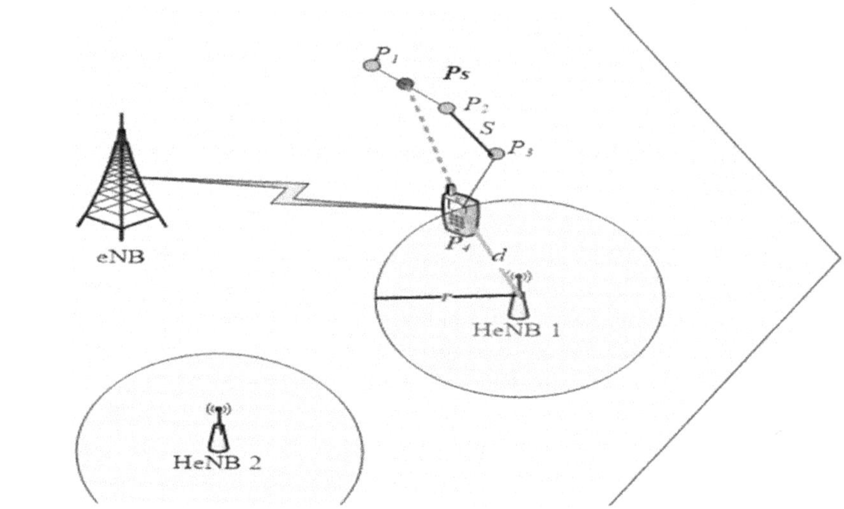

FIGURE 3.7 The eNB and HeNBs scenarios.

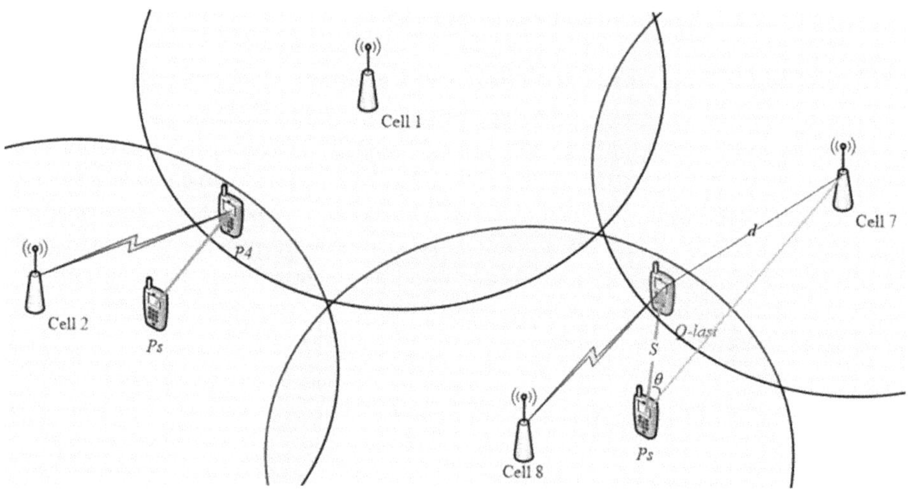

FIGURE 3.8 UE trajectory and candidate cells selection.

and to reduce the corresponding number of signaling messages exchanged between neighboring cells, optimized handover procedures are required. The serving cell is applied (Equation 3.2) for selecting the candidate cells in the handover processes. Henceforth, each neighbor cell of the UE will be verified by calculating the angle of < *Ps, Q(last), neighbor–cell–location* to be in that range; otherwise, it will be ignored, as shown in Figure 3.8. Cell 7 in Figure 3.8 is considered as the candidate cell when $\theta \le |\pm\alpha^{\circ}|$, where α is the range of the angle. In our model, and because of the use of two types of wireless (eNB and HeNB), there are also two types of cell selection ranges. In the HeNB, the coverage area is quite small; thus, the range of the angles is somehow closer to the front of the UE trajectory. Therefore, the range of the candidate cell list is $| \pm\alpha^{\circ}|$ and each neighbor HeNB cell location, which belongs to this range, will have the chance to be a candidate cell. Thereafter, the target cell is selected from the candidate cell list depending on the smallest θ and closest distance to the $UE - Q(last)$ position. The Weight Adjustment algorithm (3.1) is used to apply this technique.

In addition, our proposed model is applied in three types of handover: the hand-in, inter-HeNBs, and hand-out.

3.4.2.3.1 Hand-in Handover Procedure

In the hand-in handover scenario, the HeNBs are deployed under the eNB coverage area. Thus, each eNB knows all the HeNB_ids and locations to which the eNB belongs to. Moreover, because of the difference in the transmitting power between the two, the distance will be considered in the handover decision. As illustrated in Figure 3.9, when the UE finds the distance between itself and the neighbor cell to be less than the predefined value in the measurement control phase, the UE will send a measurement report to its serving cell (eNB), after which the eNB will start verifying each neighbor cell based on the UE direction.

The verification in Figure 3.9 depends on the θ value; if $\theta \leq |\pm\alpha^{\circ}|$, the neighbor HeNB cell will be considered in the candidate cell list. Therefore, the candidate cell list includes *HeNB*1 and *HeNB*3 based on Figure 3.9. Subsequently, the serving cell (eNB) will select the target cell, which is the *HeNB*1, from the candidate list based

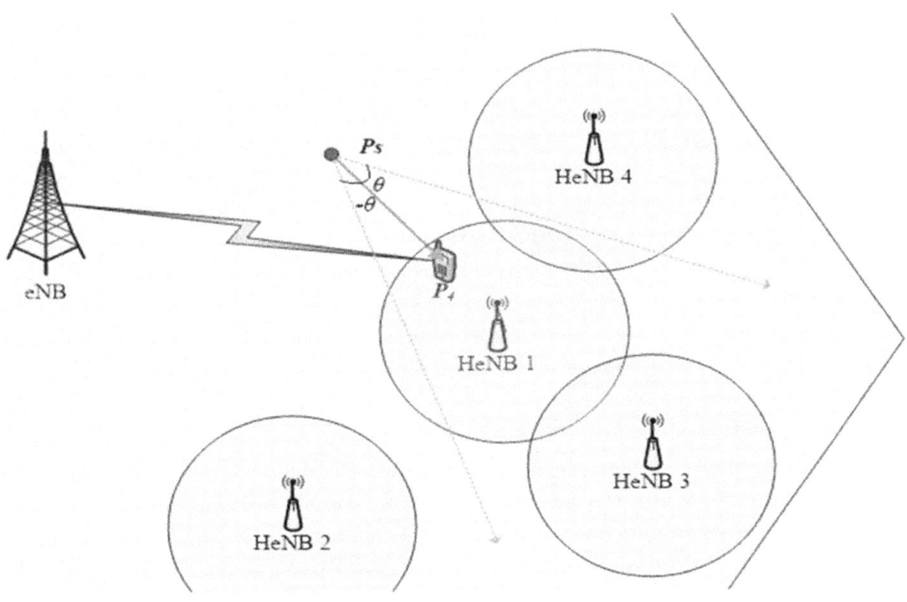

FIGURE 3.9 The hand-in handover procedure.

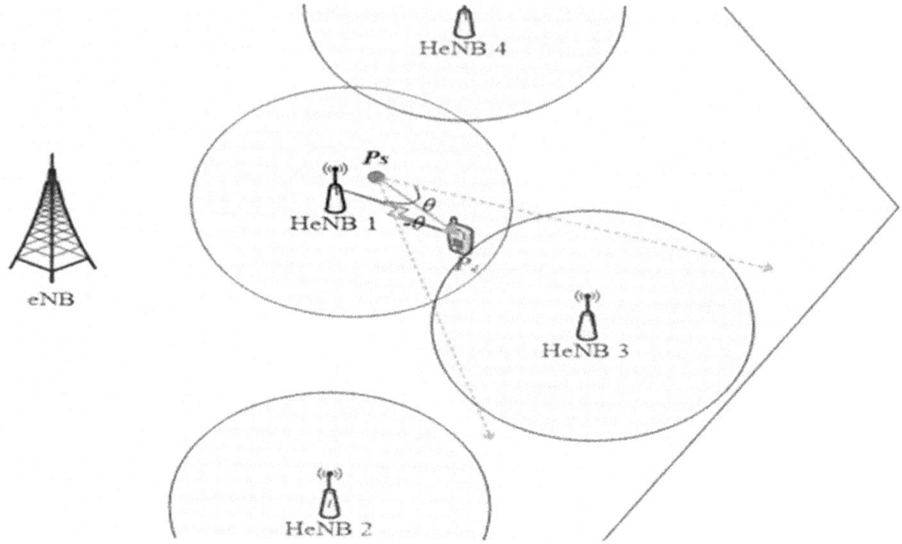

FIGURE 3.10 Inter-HeNBs handover procedure.

on the Weight Adjustment algorithm. Thereafter, the serving cell (eNB) will start to contact the target cell for the handover preparation through the X2 interface. The target cell performs admission control based on the serving cell message request, and then replies back via an acknowledged message to the serving cell, in terms of acceptance or rejection of the serving cell handover request. After that, the serving cell will inform the UE to stay connected with or to prepare the UE for the handover.

3.4.2.3.2 Inter-HeNBs Handover Procedure

In the inter-HeNBs handover scenario, the handover occurs based on the distance between the UE and neighbor HeNBs. When the distance between the UE and its serving HeNB exceeds the threshold value and the distance between the UE and neighbor HeNBs is less than the predefined value, the UE will send the measurement report message to its serving HeNB to start the handover, and as aforementioned, this message will include two points of the UE trajectory. Thereafter, the serving HeNB will select the target cell, where $\theta = |\pm\alpha^{\circ}|$. Therefore, each HeNB knows where each eNB belongs to and all HeNBs locations that are in the same group. As shown in Figure 3.10, the UE will be handed off to HeNB3.

3.4.2.3.3 Hand-out Handover Procedure

In the hand-out handover scenario, handover occurs when the UE moves toward HeNB and the distance between the UE and its serving HeNB is less than the threshold value. Moreover, there are no neighbor HeNBs that are close to the UE during its measurement control phase. Therefore, the UE will return to its serving eNB if the RSS of the eNB is still higher than the threshold value. Therefore, a variable that is called (*BSs*) is used, which is stored in the UE, to keep the *eNB– id* belonging to it.

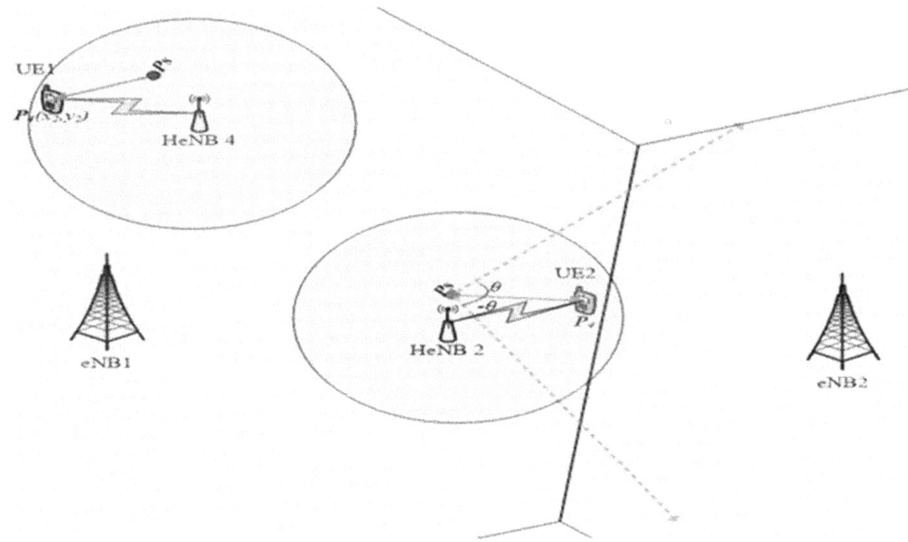

FIGURE 3.11 Hand-out handover procedure.

Thus, the UE will send the measurement report message, which includes the *BSs* value, to its serving HeNB to hand off to the main eNB. On the other hand, when the RSS between the UE and the eNBs is less than the threshold value, the UE will send a measurement report message to its serving HeNB, which includes two points of the UE trajectory. The serving HeNB will then start looking for the closest eNB, where $\theta = |\pm 60^{\circ}|$. Thereafter, the UE will move to the next eNB and update the variable *BSs* to a new $eNB - id$. In addition, another case of the hand-out handover is when the UE returns to its main eNB and there is no HeNB in its direction. If the RSS between the UE and its serving eNB is less than the threshold, the UE will send two points via a measurement report message to its serving HeNB to start searching for the next eNB, as shown in Figure 3.11.

Based on Figure 3.11, the UE1 will hand off to the eNB1 and the UE2 will hand off to the eNB2 after which its eNBs variable will be updated to eNB2. The proposed MDD-VHD scheme is presented in Figure 3.12.

3.5 SIMULATION MODEL

The simulation model of the handover in an LTE-A system is discussed in detail in this section. Additionally, the use of the model in validating the performance of MDIH-HD and MDD-VHD algorithms in both horizontal and vertical handover decisions is explained. Firstly, the simulation environment is described and the selected LTE-Sim simulator is discussed. Next, the simulation topologies for both of the proposed algorithms are highlighted. Then, the propagation characteristics are explained. Finally, the simulation setup is expressed, including the simulation parameters.

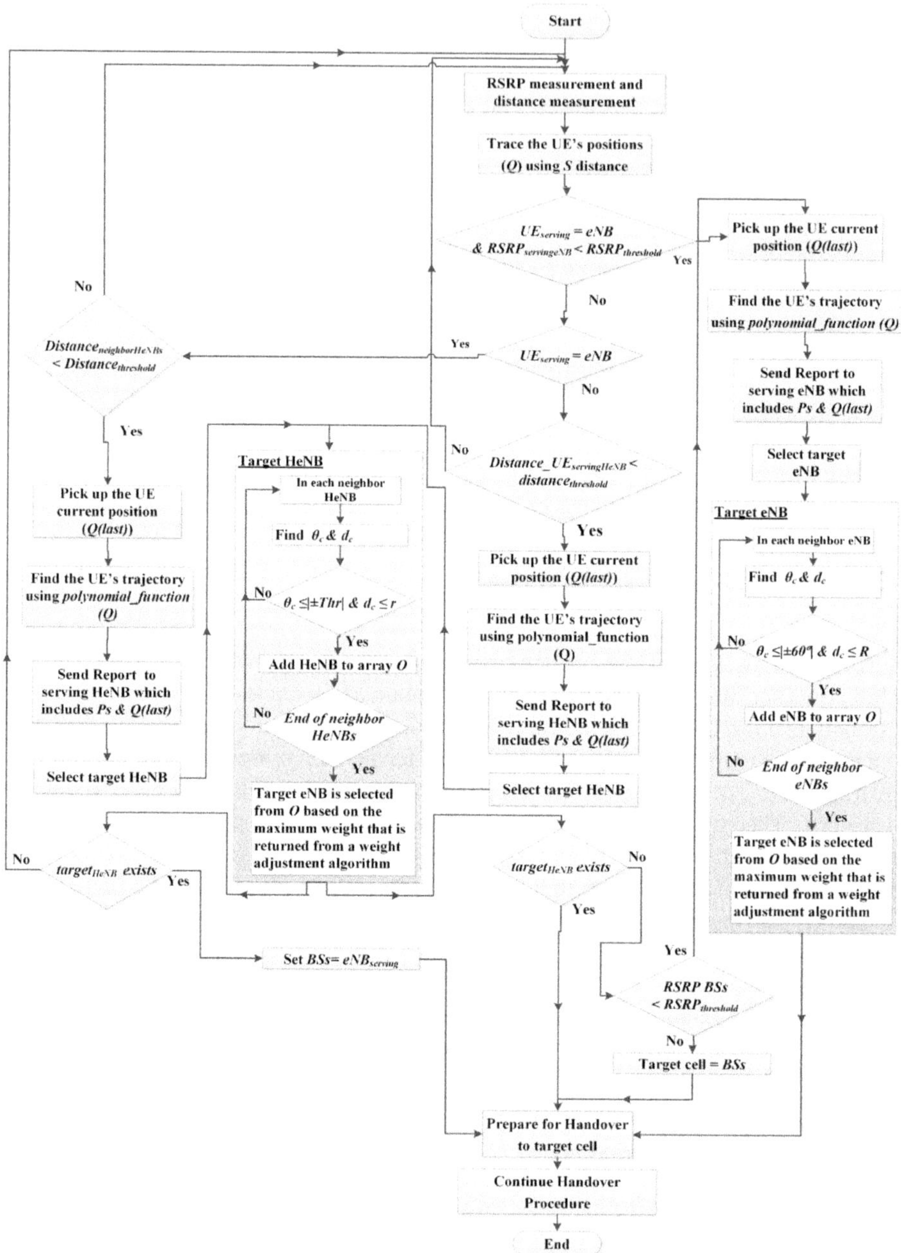

FIGURE 3.12 The MDD-VHD algorithm.

3.5.1 Simulation Tools

Performance evaluation can be done via different methods. These methods consist of theoretical analysis, computer simulation, and real-environment experiment. Computer simulation is considered to be the cheapest and easiest method. Simulations could be more helpful in creating a modeling environment for a large-scale network system rather than a practical environment (Bononi et al. 2004). Researchers can easily analyze, design, and debug a large number of UEs in a network system through computer simulations. Therefore, many network simulators can be used for the objective of this study, such as NS2 (Fall & Varadhan 2011), NS3 (S n.d.), QualNet (Scalable Network Technologies n.d.), LTE-Sim (Piro et al. 2011), WM-SIM platform (Sánchez et al. 2006), etc. Moreover, other researchers have developed their own simulators that could efficiently test and evaluate their models (Cheng-Chung Lin et al. 2012; Cheng-Chung Lin 2013).

The LTE-Sim simulator was developed with C++ by the Telematics Lab of Politecnico di Bari, Italy, as an open-source structure particularly created for simulating the LTE network system (Capozzi et al. 2012; Piro et al. 2011). LTE-Sim, considered in this book, can fulfill the requirements of this work through supporting homogeneous and heterogeneous networks in the intra-LTE-A system (eNB and HeNB). Moreover, this simulator supports all features of both E-UTRAN and EPC such as QoS management, UE velocity, mobility management, handover procedures, multi-user environment, and getaway management. Since the LTE-Sim environment does not support all the methods in this work, some improvements have been made to configure the proposed solution algorithms. Some of these LTE-Sim functions are explained in further detail in Appendix A.

3.5.2 Topology Model

In this work, there are two topologies that are concerned with benchmarking handover decision algorithms. In the MDIH-HD algorithm, the UEs move at high velocity; thus, the topology scenario needs to be drawn for a large number of eNBs and UEs. Moreover, the handover performance is examined based on different UE velocities and then benchmarked against existing works (3GPP 2013; Wang et al. 2014a). On the contrary, in the MDD-VHD algorithm, the UEs move at a low velocity. Therefore, the handover performance is examined based on the different number of HeNBs and UEs. In addition, its performance is benchmarked against existing works, such as Deswal and Singhrova (2016). The MDIH-HD and MDD-VHD topology parameters are shown in Table 3.2.

Each eNB has a radius of 1 km and consists of three sectors with no overlapping channels in between. Moreover, in MDIH-HD, UEs are distributed regularly into 19 eNBs and each set of 3 UEs is situated in the same location as the eNB, as shown in Figure 3.13(a).

The MDD-VHD topology consists of two eNBs and incremental numbers of HeNBs, which are used in each simulation scenario and configured as 20, 40, 60, 80,

TABLE 3.2
MDIH-HD and MDD-VHD topologies parameters

	MDIH-HD	MDD-VHD
	Value	**Value**
Topology Parameters	19 in an urban area	2 in an urban area
eNB number	--	20, 40, 60, 80, and 160
HeNB number	57	10, 20, 40, and 80 in each eNB
UE velocity	30, 120, and 150 km/h	25 km/h
Simulation time	110 s	130 s

Sources: Wang et al. (2014a); Deswal and Singhrova (2016); Sung et al. (2013); Wu et al. 2013.

and 160. Each HeNB radius covers 30 m. The UE number is configured as 10, 20, 40, and 80. The UEs are distributed randomly into two eNBs. Therefore, half of the UE number is present in each eNB unit and each UE starts moving from the center of its serving eNB based on random mobility.

In this study, all eNB and HeNB cells are supposed to accept the handover decision request in an environment scenario. Moreover, the handover decision in the MDD-VHD topology will cover three vertical handover types, hand-in, hand-between, and hand-out handovers, based on the availability of each vertical handover type. In addition, HeNBs will be randomly distributed in each eNB cell, for which the distribution of HeNBs will be random, and each eNB cell will be random between 50 and 1,000 m from the eNB location in three dependent scenarios (close, middle, and at the edge), as shown in Figure 3.13(b). Note that Figure 3.13 was drawn using the LTE-Sim.

3.5.3 Propagation Model

Radio propagation refers to the behavior of electromagnetic waves when transmitted from the transmitter to the receiver and their effect on the RSS of the receiver (Lee & Choi 2008). The propagation model involves path loss, shadow fading, and the fast fading models of a terrestrial cell environment.

3.5.3.1 Path Loss Model

Path loss is the decline in power density of an electromagnetic wave through expansion of radio wave in space. The path loss model, which was introduced in LTE-A, is considered in this work. The model is appropriate for scenarios in urban areas (3GPP 2016a), where the height of buildings is almost uniform. Therefore, path loss in the urban scenario can be explained using Equation (3.5):

$$L = 40.(1 - 0.004.D).\log_{10}(R) - 18.\log_{10}(D) + 21.\log_{10}(F) + 80\text{dB} \quad (3.5)$$

where R is the distance between cells and served UE in kilometers, F is the operating carrier frequency in MHz, and D is the antenna height of eNB in meters.

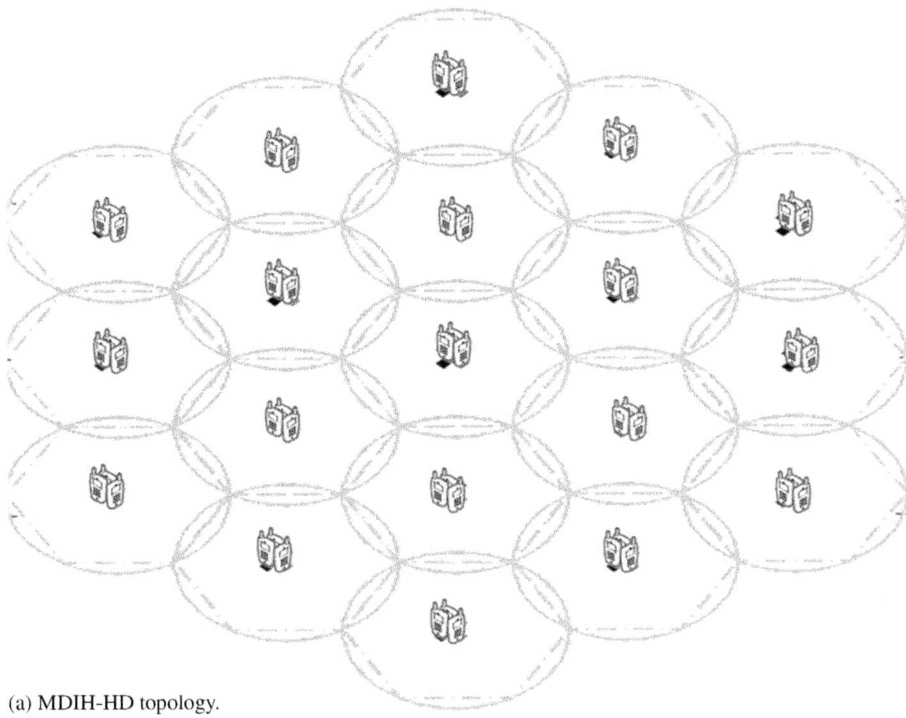

(a) MDIH-HD topology.

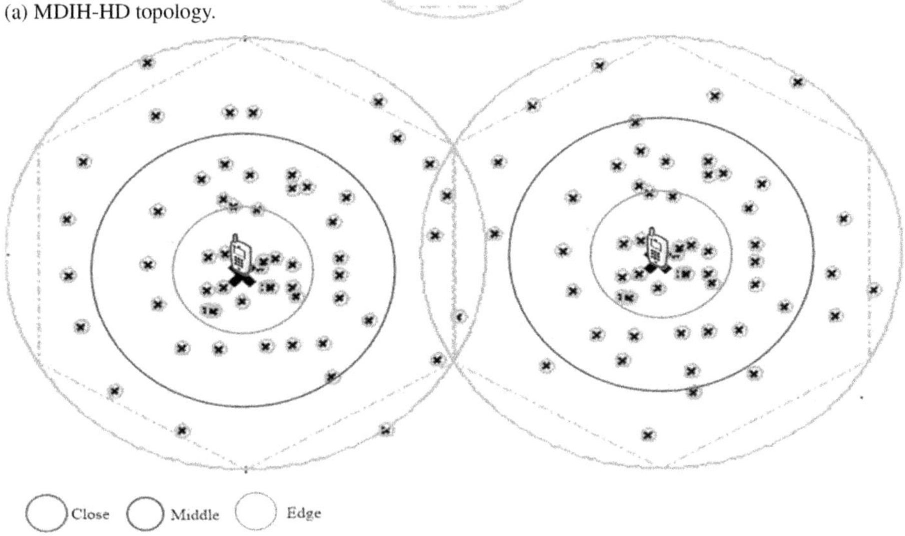

(b) MDD-VHD topology.

FIGURE 3.13 The topology models.

By considering $D = 15$ m above the average rooftop level, and a carrier frequency of 2,000 MHz (Fei & Fan 2012), the path loss model in Equation (3.5) can be simplified into Equation (3.6):

$$L = 128.1 + 37.6\log_{10}(R) \tag{3.6}$$

Additionally, the path loss for the femtocell (HeNB) introduced in LTE-A (3GPP 2010b) is also considered in this work. An HeNB urban scenario is used and the path loss for the urban scenario can be explained as per Equation (3.7):

$$L = 127 + 30\log_{10}(R) \tag{3.7}$$

3.5.3.2 Shadow Fading Model

Shadow fading indicates the difference in field strength of a radio signal, which is brought by diffraction, reflection, and the shielding phenomenon caused by trees, buildings, and mountains. The log-normal shadow fading is usually modeled as a Gaussian-distributed random variable with zero mean σ_{dB} standard deviation in dB (Goldwnamith 2005). Therefore, the shadow fading gain in this work is computed using a Gaussian-distributed random variable with a 0 dB mean and an 8 dB standard deviation (Graziosi & Santucci 2002) for eNB, and the shadowing standard deviation of HeNB signals of between 0 and 6 dB (Habibzadeh et al. 2015). The equation of shadow fading is expressed as per Equation (3.8):

$$\Psi_{\mathrm{dB}}(\omega, \mu_{\mathrm{dB}}) = \frac{\xi}{\omega.\sigma_{\mathrm{dB}}.\sqrt{2\pi}} \mathrm{e}^{-\left[\frac{(10\log_{10}(\omega)-\mu_{\mathrm{dB}})^2}{2\sigma_{\mathrm{dB}}^2}\right]} \tag{3.8}$$

where $\xi = \frac{10}{\ln 10}$, μ_{dB} is the mean value of received signal power in dB, σdB is the standard deviation of ωdBm in dB, and ω represents the transmitted-to-received power ratio $\omega = \frac{P_t}{P}$ which is included as default values in our simulation by LTE-Sim.

3.5.3.3 Fast Fading Model

The largely recognizable model used for fast fading is the Rayleigh fading model, which is applicable in multipath fading when a non-line-of-sight (NLOS) path between the transmitter and receiver is encountered. The Rayleigh model probability is given by Belloni (2004) and Patzold et al. (1996) as per Equation (3.9):

$$\begin{cases} \frac{A_{p_r}}{E^2} e^{-\left(\frac{A_{Pr}^2}{2E^2}\right)} & 0 \le p_r \le \infty \\ 0 & p_r < 0 \end{cases} \tag{3.9}$$

where A_{p_r} is the amplitude of the received signal power and $2E^2$ is the mean power of multipath received signal power (e.g., the received signal power based on path loss and shadowing alone). However, the Rayleigh model probability is included in our simulation via the LTE-Sim as a default value.

3.5.3.4 Signal to Interference plus Noise Ratio (SINR)

SINR is commonly used in a wireless system to measure the quality of wireless connections. Precise SINR approximation provides a more effective system and a higher user-perceived QoS (Khattab & Elsayed 2006). The SINR is given by Tabia et al. (2012) as per Equation (3.10):

$$SINR^{n}_{c,u} = \frac{P^{n}_{c,m} \cdot G^{n}_{c,u}}{P_{noise} + P_{i}} \tag{3.10}$$

where u is the index of the UE, c is the index of the cell, n is the index of the PRB, P is the transmission power of the serving base station, G is the channel gain between the user and the cell, and P_{noise} is the variance of the additive white Gaussian noise (3GPP 2016a). P_i is the sum of the powers received by the UE, coming from the neighbor cells, except to the one it is connected to.

3.5.3.5 Reference Signal Received Power

RSRP refers to the RSS without thermal noise and inter-cell interference. In 3GPP, the RSRP is described as the average of the linear power over the power contributions (in watt) of the PRBs (3GPP 2015b; 3GPP 2016b). RSRP can be calculated from the serving cell transmit power (P), the path loss value from UE to the serving cell (L_u), and additional shadow fading with a log-normal distribution and a standard deviation (L_{fad}) (Yang et al. 2015). The RSRP is calculated using Equation (3.11) in dB:

$$RSRP_{s,u} = P_s - L_u - L_{fad} \tag{3.11}$$

where s is the serving cell and u is the index of UE.

3.5.4 Simulation Setup

In this work, the data traffic applications which were used in the simulation are the Constant Bit Rate (CBR) and Voice over Internet Protocol (VoIP).

1. CBR application: In CBR, data packets are sent in a fixed size and fixed interval for each data packet. In the two different proposed algorithm scenarios, the CBR is used to evaluate the performance of each proposed algorithm. In both of the proposed algorithms, the CBR sources send 512-byte packets at the rate of 8 packets/second.
2. VoIP applications: VoIP can be briefly defined as transmitting the voice conversation in packets and transporting them through a packet switched network. The codec G.729 is used as a vocoder of the VoIP, the User Datagram Protocol (UDP) header size is 20 bytes, the Real-time Transport Protocol (RTP) header size is 12 bytes, and the payload size is 32 bytes. Hence, the packet interval time = 20 ms (Piro et al. 2011).

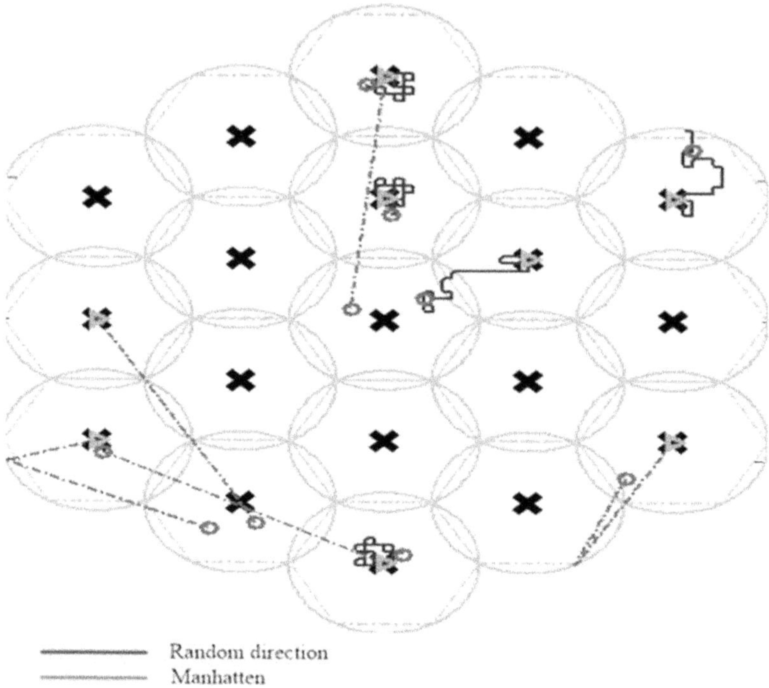

FIGURE 3.14 A sample of the UE movement inside eNBs.

In the mobility model, the LTE-Sim supports two types of UE mobility, random direction and Manhattan, as illustrated in Figure 3.14.

In Figure 3.14, the red line represents the random direction mobility model, whereas the blue line shows the Manhattan mobility model. In this work, the random direction model is selected for the UE mobility. In the random direction, the UEs move inside multi-eNBs or HeNBs, and each UE is assigned a random direction at the beginning of its data session and moves within the cell at a constant velocity . In addition, when the UE reaches the border of the topology model, the UE goes back inside and completes its movement until the simulation time is terminated.

Moreover, the LTE-A system is modeled according to 3GPP specifications (3GPP 2016a; 3GPP 2016c) and a maximum delay is the value that the data packet is considered to be too late if this period exceeds. Table 3.3 summarizes the LTE-Sim parameters.

3.6 PERFORMANCE METRICS

As previously discussed in the subsection of the proposed handover solutions, the outcome of the proposed algorithms is measured using units known as performance metrics. With regard to the performance evaluation metrics, the handover number, signaling measurement number, packet delay, packet loss, and network throughput are selected in the study's investigation into the effectiveness of the proposed

TABLE 3.3
The simulation parameters

Simulation Parameters	Value
Environment	NB and HeNB, (Urban areas), LTE-A
Frequency	2 GHz
Bandwidth	5 MHz
Noise density	–174 dBm/Hz
Subcarrier spacing	15 kHz
Resource block	180 KHz
eNB transmit power	43 dBm
eNB transmit power	21.5 dB
Radius of each eNB	1 km
Radius of each HeNB	30 m
NB Path loss	$128.1 + 37.6 \log_{10}(R)$
HeNB Path loss	$127 + 30log_{10}(R)$
Data Traffic	CBR & VoIP
Mobility model	Random direction
eNB Penetration loss	10 dBm
HeNB Penetration loss	0dBm
HeNB minimum coupling loss	43 dBm
Handover parameters	
eNB $RSRP_{threshold}$	–80 dBm
HeNB $RSRP_{threshold}$	–74 dBm
TTI	8 ms
TTT	10 ms
RSRP measurement period	10 ms
Maximum delay	10 ms

Sources: Deswal and Singhrova (2016); Xu et al. (2013); 3GPP (2010a); 3GPP (2012d); Munoz et al. (2013); Piro et al. (2011); Wang et al. (2014a); Capozzi et al. (2012); Chen et al. (2014a); Habibzadeh et al. (2015).

handover decision algorithms. Besides that, the new proposed algorithms concentrate on improving the performance of the handover procedure through reducing the number of unnecessary handovers and increasing data packet delivery in minimum delay and loss. Furthermore, the algorithms also aim to decrease the number of signaling measurement messages, which in turn minimize the UE battery consumption. In addition, with the reduction of the above factors, the network system throughput will be increased. Therefore, the following performance metrics are selected to assess these aspects in the proposed handover algorithms that are repeatedly sent.

3.6.1 Average of Handover Number

It is the average of handing over the UE from the serving to the target cells once the target cell signal quality goes better than the serving cell quality per second. Therefore, the reduction in handover number minimizes the handover failure and increases the QoS, as each handover process is initiated by hazard failure. The average number of handovers (ANH) is calculated by Equation (3.12).

$$ANH = \frac{N_{HO}}{T} \tag{3.12}$$

where N_{HO} is the total number of handovers and T is the simulation time.

3.6.2 Average of Signaling Measurements Number

It's the UE scans to examine the RSS of neighbor cells to select the target in every second. In standard handover, the signaling measurement is executed periodically to examine the neighbor cells. Therefore, the system power is wasted during signal measurements; thus, the reduction in signaling measurements number will save the UE's power. The Average of Signaling Measurement Number (ASMN) is calculated by Equation (3.13).

$$ASMN = \frac{N_{SM}}{T} \tag{3.13}$$

where N_{SM} is the total number of signaling measurements.

3.6.3 Packet Loss Ratio

This metric evaluates the real-time service because retransmission is not applicable. The invalid packets are defined as those arriving at their destination in a delay more than 100 ms (3GPP 2015a). The Packet Loss Ratio (PLR) is calculated by Equation (3.14).

$$PLR = \frac{Pack_s - Pack_r}{T} \tag{3.14}$$

where $Pack_s$ is the total number of packets that is created through simulation time and $Pack_r$ is the total number of received packets.

3.6.4 Packet Delay Ratio

It is defined as the time period of the data packet departure time and its arrival time at the buffer of the cell. The Packet Delay Ratio (PDR) is calculated by Equation (3.15).

$$PDR = \frac{\sum_{k=1}^{N} \sum_{i=1}^{u} delay_{ki}}{E} \tag{3.15}$$

where $delay_{ki}$ is the total packet delay of UE_i at its serving $cell_k$, u is the total number of UEs, and E is the total number of packets that all of the UEs received during simulation time. N is the total number of cells.

3.6.5 Throughput

It is defined as the rate of data packets delivered successfully to the destination over a certain time duration in bits per second. The system throughput (TH) is calculated by Equation (3.16).

$$TH = \frac{\sum_{k=1}^{N} \sum_{i=1}^{u} TH_{ki}}{T} \tag{3.16}$$

where the TH_{ki} is the total size of packet correctly received by UE_i.

3.7 SIMULATION PARAMETER SETTING

In this section, the input parameter threshold values are validated to select the best threshold values in both proposed algorithms.

3.7.1 The MDIH-HD Input Parameters Setting

The MDIH-HD algorithm was proposed to enhance the handover decision procedure in horizontal networks (eNBs) when the UE moves at high velocity. The MDIH-HD algorithm depends on user profile and user information history. In order to enhance the efficiency of the handover procedure, β_{Thd} and P_{Thd} threshold values are considered, allowing the UE to have a little more free space away from its previous trajectory during matching of the current UE trajectory with the data in the UE table. Moreover, the S_{Thd} threshold value is considered in tracing the UE positions for every S distance. The experimental results were applied to the MDIH-HD algorithm in LTE-Sim to tune the β_{Thd}, P_{Thd}, and S_{Thd} parameters through comparing between their throughputs and then choosing the highest one. The simulation parameters, which were discussed in Section 3.5, have been applied in MDIH-HD to select the best throughput results in the case when the UE velocity is 120 km/h for the CBR application. Based on Figure 3.15, the results show that the best system throughput is achieved when the angle (β_{Thd}) between the current UE trajectory and the trajectory in the UE table is 15°. The throughput shows an increase in value when β_{Thd} starts increasing from 5° to 15° after which it decreases when β_{Thd} is 20°. However, the throughput results increase again when β_{Thd} are 25° and 30° but this increase is lesser than the throughput increase at a β_{Thd} is 15°.

In Figure 3.16, the results show that the best system throughput is achieved when the distance (P_{Thd}) between the current UE position and the position in the UE table is 20 m. The throughput increases when P_{Thd} increases from 10 to 20 m, after which it decreases when P_{Thd} increases from 25 to 30 m. This P_{Thd} threshold value gives the

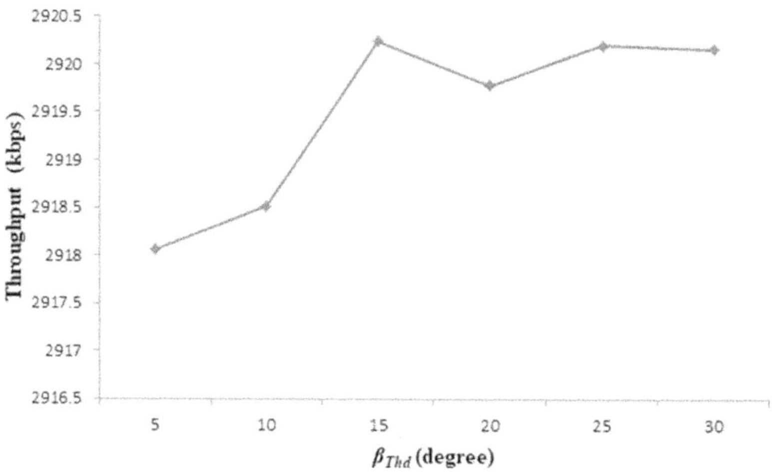

FIGURE 3.15 The MDIH-HD algorithm for different β_{Thd} values.

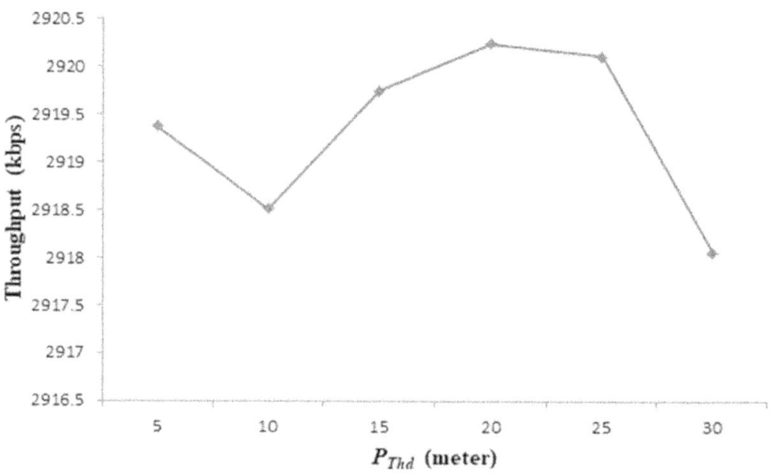

FIGURE 3.16 The MDIH-HD algorithm for different P_{Thd} values.

UE a little more free space to select the nearest current UE position from the UE table if available.

Figure 3.17 shows that the best distance value (S_{Thd}) for the UE tracking positions is 150 m in the MDIH-HD algorithm. Moreover, Figure 3.17 also shows a decrease in throughput results when S_{Thd} is less or more than 150 m.

By comparing the MDIH-HD algorithm to the 3GPP and the handover algorithms in Wang et al. (2014a), the two benchmark handover algorithms are observed to use their own parameters. In Wang (2014), the handover algorithm is used when UE picks up the third position and then finds the angle between the last three positions to be less than the predefined threshold value and finds RSS of the neighbor cells to be better

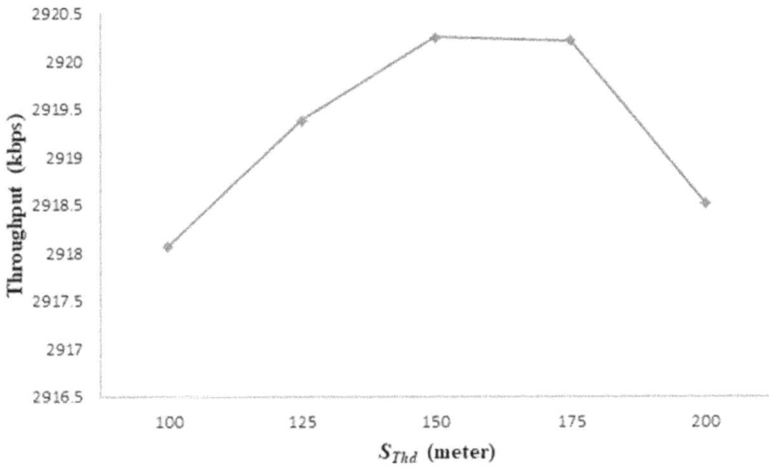

FIGURE 3.17 MDIH-HD algorithm for different S values.

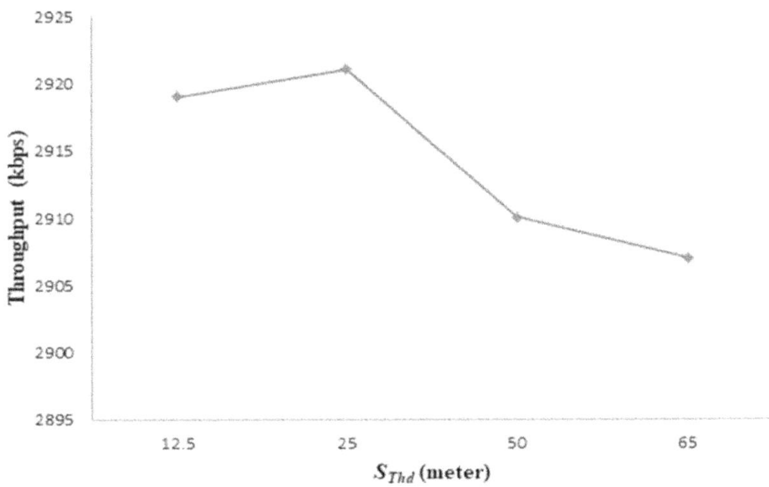

FIGURE 3.18 The algorithm of Wang (2014) for different S values.

than its serving cell. Therefore, the handover occurs even when the RSS between the UE and its serving cell is still good. On the other hand, if the handover is required and the angle between the last three UE positions is bigger than the predefined value, the LTE\LTE-A standard handover will be used and their proposed algorithm is ignored. Thus, in Wang (2014), the parameters were tuned by comparing their performance and selecting the best parameter value. The best distance (S) between the three positions of UE for the best performance in the algorithm of Wang (2014) is 25 m, as shown in Figure 3.18.

3.7.2 The MDD-VHD Input Parameters Setting

In this section, the simulation topology parameters of the MDD-VHD algorithm discussed in Section 3.5 are used to investigate and validate the threshold values of S, $HeNB{-}d_{threshold}$, and θ parameters through experimental simulation results. In addition, the best duration time for the simulation is found to be 130 s. This condition allows the UE with a speed of 25 km/h to execute the handover among cells, as the UE starts moving from its eNB tower location. Moreover, the HeNB numbers are distributed randomly at a distance of 50–1,000 m from the eNB location, in which the HeNB is the open access type, which accepts any handover request. In addition, data traffic application is configured as CBR, the UE number is configured as 10, and the HeNB number is configured as 40.

The experimental results were applied to the MDD-VHD algorithm in the LTE-Sim to tune the S, $HeNB{-}d_{threshold}$, and θ parameters through a comparison of their performance. Figure 3.19 shows that the experimental distance value (S) between the previous and the current position of UE is 2 m.

Based on Figure 3.19, the highest throughput result for the S parameter in the MDD-VHD algorithm is 2 m. On the other hand, Figure 3.19 also shows a decrease in the throughput results when S is less or more than 2 m.

The experimental results of the $HeNB{-}d_{threshold}$ threshold value, which allows the UE to hand off either from eNB to HeNB, HeNB to HeNB, or from HeNB to eNB, are illustrated in Figure 3.20. Based on Figure 3.20, the best $d_{threshold}$ value is selected according to the best throughput result that intersects with the highest handover number. The first value of the $d_{threshold}$ is 27 m, showing the best performance but the lowest number of handovers.

This means that the UE could not connect to the HeNB until it is at very close to the HeNB. The second $d_{threshold}$ experimental value yields the best handover number and a good throughput value, which means that the UEs could connect to the HeNBs

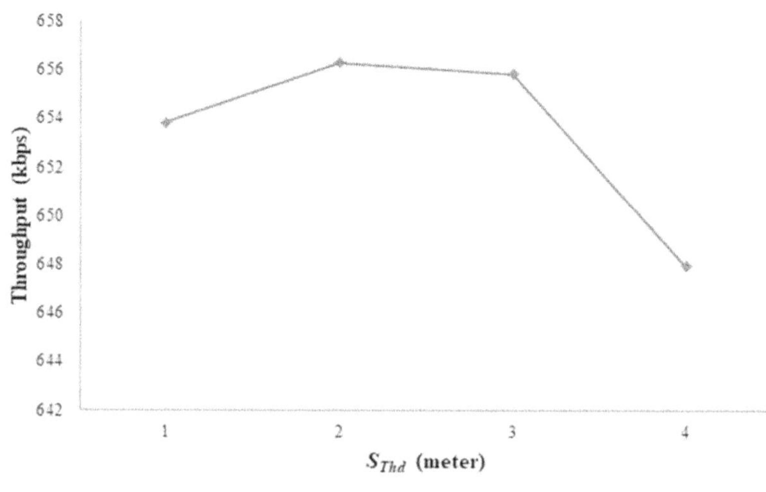

FIGURE 3.19 MDD-VHD throughput performance for different S values.

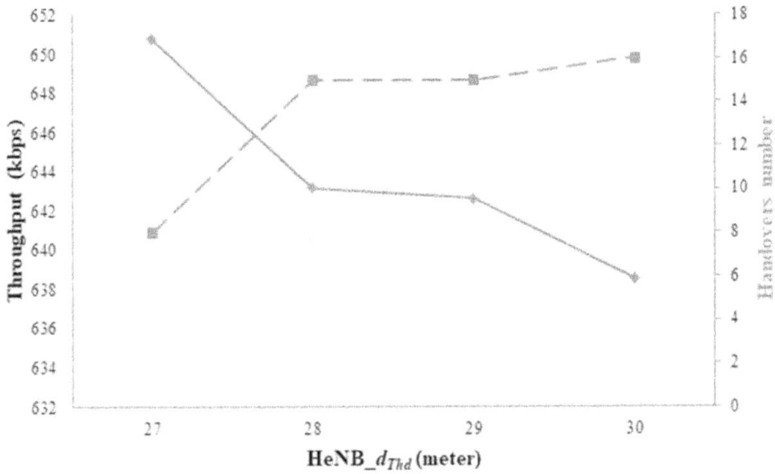

FIGURE 3.20 MDD-VHD throughput and handovers number for different *HeNB* ^{d}*threshold* values.

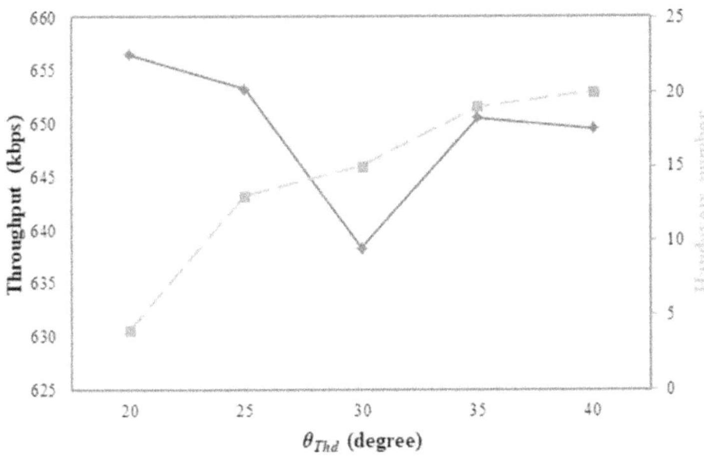

FIGURE 3.21 MDD-VHD throughput and handovers number for different θ parameter values.

without the need to move closer to the HeNB tower. This is in contrast to the first $d_{threshold}$ value. The third and fourth $d_{\text{threshold}}$ experimental values show a decrease in throughput and an increase in handover number, and thus will be ignored. Therefore, based on the first and second $d_{threshold}$ experimental values, the distance of 28 meters was selected as the HeNB distance threshold value.

With regard to the θ threshold value, which is used in searching for the target HeNB in front of the UE direction, the experimental threshold value is presented in Figure 3.21. The value of θ should not be too large to prevent UE from connecting with the neighbor HeNBs in a short time, as the HeNB is not completely ahead of the current UE position during the handover procedure time. On the other hand, the θ parameter should not be close to zero to allow the UE to connect to the HeNB, which is closer to or in front of the current UE position.

Based on Figure 3.21, the θ threshold value was selected from the experimental θ parameter value, depending on the highest performance and the highest handover number.

The first value of θ parameter is 20°, which yields the highest throughput, but the lowest handover number. These results are predictable because of the small angle used. The second value of the θ parameter shows the best performance with good results for the handover number. Moreover, for the third θ parameter, which is 30°, the highest handover number but the throughput result is low; thus, this θ value will be ignored. For the last two θ parameter values, the figure shows lower performance results and higher handovers number compared to the second θ values. Therefore, the maximum cosine function (θ) value which is selected in this study is$| \pm 25°|$.

3.8 SUMMARY

This chapter describes the generic design of the approach for enhancements used in the simulator development of this study. The concept of proposed solutions, simulation model, and performance evaluation are discussed and analyzed to guarantee that the research objectives can be carried out successfully. The proposed solution phase describes the proposed handover algorithms in the LTE-A system regarding both homogeneous and heterogeneous networks. In addition, it also describes the integration between the two proposed handover decision algorithms. Moreover, the proposed simulation model is discussed in detail. The propagation model and the simulation setup are also explained. Finally, handover performance metrics are illustrated and the input parameters setting for each algorithm are discussed in detail. The proposed MDIH-HD algorithm in enhancing handover performance in high-speed mobility is presented in Chapter 4. Chapter 5 discusses the proposed MDD-VHD algorithm in enhancing handover performance in low-velocity mobility.

4 Moving Direction Informative Historical-Handover Decision Algorithm

4.1 INTRODUCTION

The new MDIH-HD is proposed based on UE trajectory and UE information history to improve the horizontal handover decision in LTE-A. The performance of the proposed MDIH-HD is presented and discussed in this chapter. The simulation uses three different UE velocities (30, 120, and 150 km/h) in two traffic applications (CBR and VoIP). The performance of the MDIH-HD is measured at every UE velocity and in each application according to five performance metrics, which are the average number of handovers, the average signaling measurement number, packet delay, packet loss, and throughput. The performance of the MDIH-HD algorithm is compared to the 3GPP standard handover decision and the handover decision in Wang et al. (2014a). Finally, the results of the proposed algorithms are presented and discussed, and the conclusion remarks are given.

4.2 PERFORMANCE EVALUATION OF MDIH-HD ALGORITHM

The effectiveness of the MDIH-HD algorithm and the other two benchmark handover decision algorithms was evaluated using the LTE-sim simulator with the same network configuration, as described in Chapter 3. Moreover, the MDIH-HD algorithm was executed two times in order to fill each UE table and thus obtain UE table proficiency. AT every UE velocity, five performance metrics are measured, which are the average handover number, average signaling measurement rate, packet delay rate, packet loss rate, and throughput. The performance comparison and an in-depth analysis of the results at every UE velocity are presented in the following sections. Moreover, the simulation work was carried out through integrating the LTE-Sim simulators; each simulation was carried out ten times in order to obtain average result values. The simulation results and analysis based on these metrics are displayed and analyzed in the following subsections.

DOI: 10.1201/9781003648130-4

4.2.1 Average Handover Number

The results of average handover number for the proposed MDIH-HD algorithm and the other two benchmarking handover decision algorithms at different high UE velocities (30, 120, and 150 km/h) are presented in Figure 4.1.

Based on each result, it is obvious that by increasing the UE velocities, the algorithms show increment in the average handovers number, because the UEs move in a long distance. The MDIH-HD increases from 0.0031 to 0.0177 when the UE velocity increases from 30 km/h to 150 km/h, while the Wang algorithm increases from 0.0036 to 0.0181 with the same change in velocities. In addition, there is an increase in 3GPP handover decision algorithm from 0.0044 to 0.0213 with the same change in velocities.

The results emphasize that the best performance was achieved by the MDIH-HD algorithm for all UE velocities. That is because the handover decision of the proposed algorithm is not triggered until the RSRP between the UE and its serving eNB is less than the predefined threshold value, and this contributes to avoiding unnecessary handover. Moreover, when UE velocity is 30 km/h, the total number of handovers is read from the UEs tables. On the other hand, if the UE velocity is 120 km/h, 81% of handovers are read from the UE table and the rest will be read using the moving direction prediction procedure. Finally, if the UE velocity is 150 km/h, 65% of handovers are read from the UE table and the moving direction prediction procedure is applied on the rest. Contrariwise, in the case of Wang and 3GPP standard, the handover decision is triggered when the RSRP between the UE and its neighbor eNBs is higher than the RSRP between the UE and its serving eNB. Moreover, Wang algorithm reduces the average number of handovers more than the 3GPP standard algorithm through

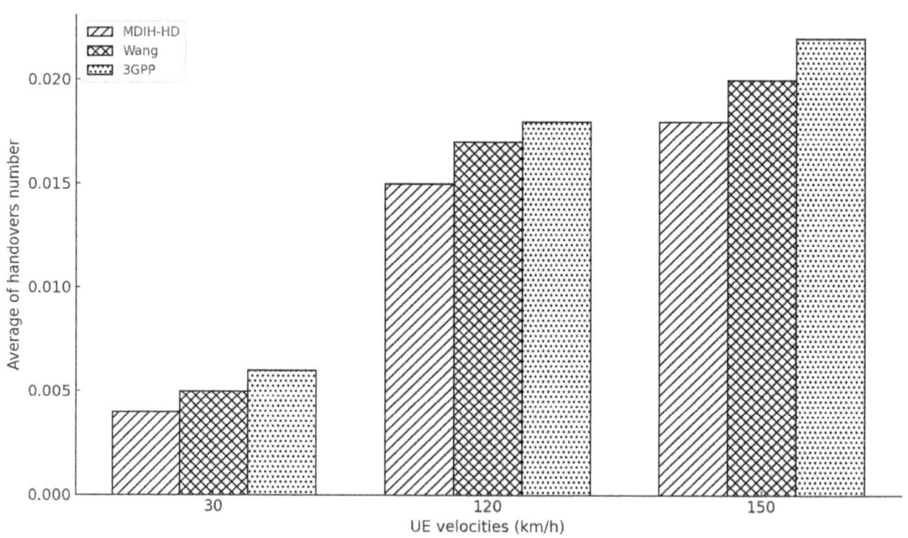

FIGURE 4.1 Comparison of average handovers number.

TABLE 4.1
The percentage of MDIH-HD reduction in average of handovers number compared to Wang and 3GPP algorithms

UE Velocity	Reduction Percentage Compared to Wang	Reduction Percentage Compared to 3GPP
30	13%	28%
120	6.5%	9%
150	2.6%	17.3%

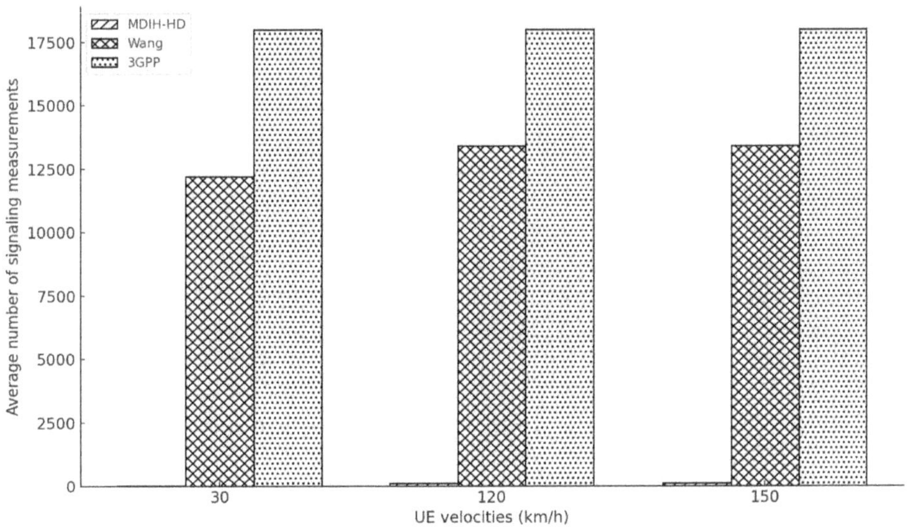

FIGURE 4.2 Comparison of signaling measurements number.

selecting the target eNB which exists in front of UEs if the UEs move in the same direction of their trajectories.

However, the proposed MDIH-HD algorithm reduces the average number of handovers in comparison to Wang and 3GPP handover algorithms at every UE velocity as presented in Table 4.1.

Finally, concerning the total average number of handovers, it is reduced by the MDIH-HD algorithm by 7.3% compared to Wang algorithm and by 18% compared to 3GPP handover algorithm for all different UE velocities (30, 120, and 150 km/h).

4.2.2 Average Signaling Measurements Number

By increasing the UE velocity from 30 to 150 km/h, as shown in Figure 4.2, the MDIH-HD algorithm shows a sharp decrease in the average signaling measurements rate. It was noted that the average number of signaling measurements in the 3GPP standard handover scheme is not affected by the change in UEs velocity. Thus, the 3GPP standard handover procedure proved that the duration of signaling measurement is fixed. Moreover, the 3GPP handover decision applies the signaling measurement control on all UE neighbor eNBs in each measured period through comparing the RSRP of the UE with its serving eNB on one hand, and with UE neighbor eNBs on the other hand. In Wang handover decision, the signaling measurements control is applied on the eNBs which are located ahead of UE trajectory if the UE moves in the same pathway. On the contrary, if the UE movement trajectory is not in the same pathway and the UE needs to apply a handover, the 3GPP handover decision will be applied on Wang handover decision procedure.

Nevertheless, our proposed scheme reduced the number of signaling measurements sharply compared to 3GPP standard and Wang algorithms. The reason for this is the use of UE trajectory table which allows the UE to do signaling measurements only with the target eNB. Moreover, in the case when UE trajectory does not exist in the UE table, the serving eNB executes the moving direction prediction procedure based on UE direction, UE positions, and the distance between the UE and the neighbor eNBs in order to choose the target eNB taking into account that the eNB coverage area is considered to be 1 km.

The average signaling measurements number in MDIH-HD increases from 20 to 111 when the UE velocity increases from 30 to 150 km/h, while in Wang algorithm the average signaling measurements number increases from 12,212 to 13,412.18 with the same change in UE velocity. In addition, the average signaling measurements number is stable at 17,999.845 for all UE velocities in the 3GPP handover.

The average signaling measurements number is reduced by more than 99% in both 3GPP standard and Wang schemes, which means a reduction in power consuming.

4.2.3 Packet Delay Rate

With regard to the packet delay rate, the MDIH-HD algorithm is evaluated using various high velocities of UE (30, 120, and 150 km/h) in two applications (CBR and VoIP). Figure 4.3 shows the packet delay rate in both applications (CBR, VoIP, and both) in the MDIH-HD algorithm.

In all the applications, Figure 4.3 shows an increase in the packet delay rate because of the effect of the mobility velocity on the handover decision. The CBR application shows the highest packet delay rate because of the packet size which is 512 bytes with the rate of 8 packets/second. The packet delay increases from 0.0024 to 0.0032 s when the UE velocity increases from 30 to 150 km/h. Regarding the VoIP application, the same figure shows that it has the lowest packet delay rate because of the packet size and priority. The total packet size for the VoIP is 32 bytes in 20 ms interval time. The packet delay increases from 0.0014 to 0.0019 s when the UE velocity increases from 30 to 150 km/h. Moreover, when applying both applications together in the same scenario, the figure shows the medium result of the packet delay rate and that is because the packets size is divided between the two applications. The packet delay increases from 0.0019 to 0.0022 s when the UE velocity increases from 30 to 150 km/h.

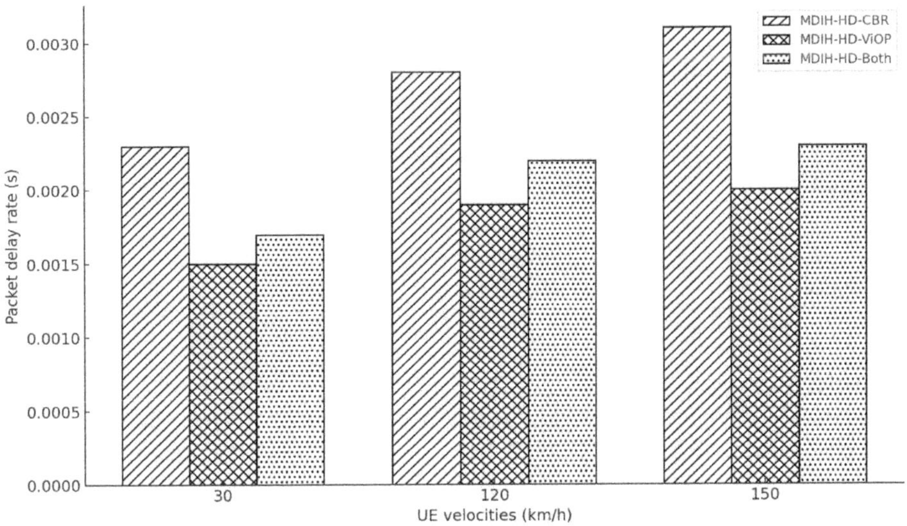

FIGURE 4.3 Comparison of packet delay rate between applications in MDIH-HD algorithm.

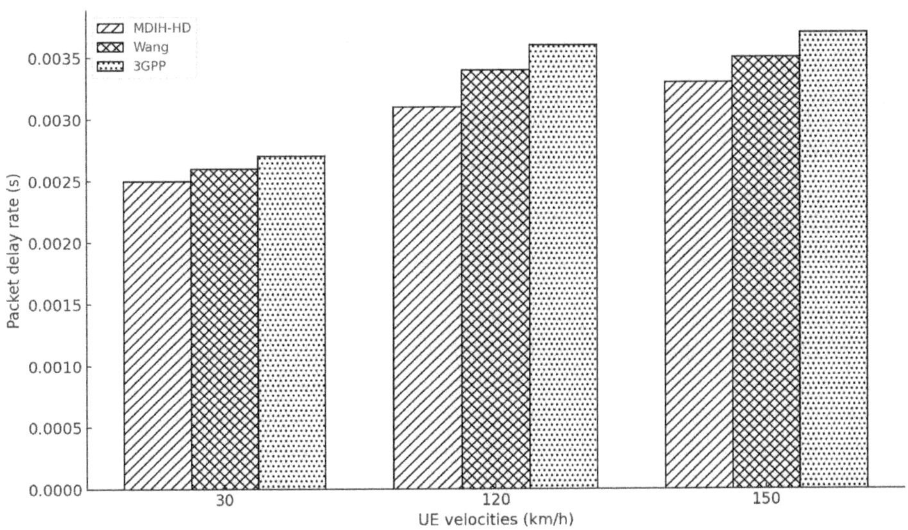

FIGURE 4.4 Comparison of packet delay rate using CBR application in three handover algorithms.

Regarding the CBR application in the three handover decision algorithms using different high UE velocities (30, 120, and 150 km/h), Figure 4.4 shows that all handover decision algorithms increment the packet delay rate with an increase in the UE velocity because of the effect of the velocity on the handover decision procedure. The handover at high-velocity mobility needs to be faster and the delay in selecting the

TABLE 4.2
The percentage of MDIH-HD reduction in packet delay rate compared to Wang and 3GPP algorithms in CBR application

UE Velocity	Reduction Percentage Compared to Wang	Reduction Percentage Compared to 3GPP
30	3.6%	6.6%
120	10%	18.7%
150	6.8%	12.1%

target eNB during the searching time will increase the packet delay rate. The MDIH-HD increases the packet delay rate from 0.0024 to 0.0032 s when the UE velocity increases from 30 to 150 km/h, while in Wang the packet delay rate increases from 0.0024 to 0.0035 s with the same change in UE velocity. Moreover, the packet delay rate increases in the 3GPP handover algorithm from 0.0025 to 0.0037 s with the same change in UE velocity. However, the results show that the MDIH-HD algorithm is more efficient at high velocities compared to Wang and 3GPP handover algorithms.

However, the proposed MDIH-HD algorithm reduces the packet delay rate compared to Wang and 3GPP handover algorithms in CBR application at every UE velocity as presented in Table 4.2.

Finally, concerning the packet delay rate in CBR application, it is reduced in the MDIH-HD algorithm by 6.8% compared to Wang algorithm and by 12.4% compared to 3GPP handover algorithm for all different UE velocities (30, 120, and 150 km/h).

With regard to VoIP application in the three handover decision algorithms using different high UE velocities (30, 120, and 150 km/h), Figure 4.5 shows that all handover decision algorithms increment the packet delay rate with an increase in the UE velocity.

The MDIH-HD increases the packet delay rate from 0.0014 to 0.0018 second when the UE velocity increases from 30 to 150 km/h, while in Wang the packet delay rate increases from 0.0017 to 0.0022 s with the same change in UE velocity. Moreover, the packet delay rate increases in 3GPP handover algorithm from 0.0018 to 0.0022 s with the same change in UE velocity. Nevertheless, with respect to VoIP application, the packet delay rate has the lowest results in all handover algorithms for the three mobility velocities compared to the CBR application due to the different packet size and priority in both applications.

However, the proposed MDIH-HD algorithm reduces the packet delay rate compared to Wang and 3GPP handover algorithms in VoIP application at every UE velocity as presented in Table 4.3.

Finally, concerning the packet delay rate in VoIP application, it is reduced in the MDIH-HD algorithm by 11.3% compared to Wang algorithm and by 14.9% compared to 3GPP handover algorithm for all different UE velocities (30, 120, and 150 km/h).

In addition, when both applications were applied in the same scenario using the three handover decision algorithms at different high UE velocities (30, 120, and 150 km/h), the results are shown in Figure 4.6.

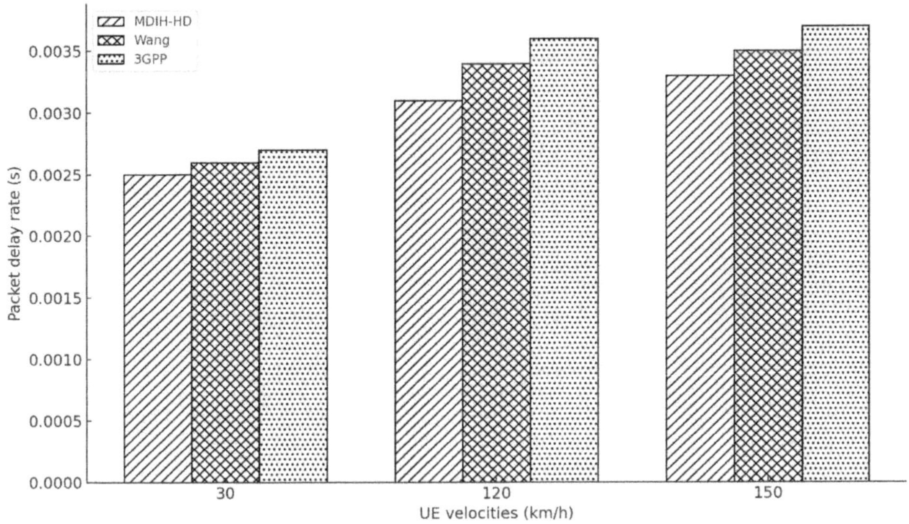

FIGURE 4.5 Comparison of packet delay rate using VoIP application in three handover algorithms.

TABLE 4.3
The percentage of MDIH-HD reduction in packet delay rate compared to Wang and 3GPP algorithms in VoIP application

UE Velocity	Reduction Percentage Compared to Wang	Reduction Percentage Compared to 3GPP
30	15.3%	18.1%
120	3.9%	11%
150	14.8%	15.6%

Based on Figure 4.6, the MDIH-HD increases the packet delay rate from 0.0019 to 0.0021 s when the UE velocity increases from 30 to 150 km/h, while in Wang the packet delay rate increases from 0.0020 to 0.0027 s with the same change in UE velocity. Moreover, the packet delay rate increases in 3GPP handover algorithm from 0.0020 to 0.0040 s with the same change in UE velocity.

However, the proposed MDIH-HD algorithm reduces the packet delay rate in comparison to Wang and 3GPP handover algorithms in both applications at every UE velocity as demonstrated by Table 4.4.

Finally, concerning the packet delay rate in both applications, it is reduced in the MDIH-HD algorithm by 12.7% compared to Wang algorithm, and by 24.4% compared to 3GPP handover algorithm for all different UE velocities (30, 120, and 150 km/h).

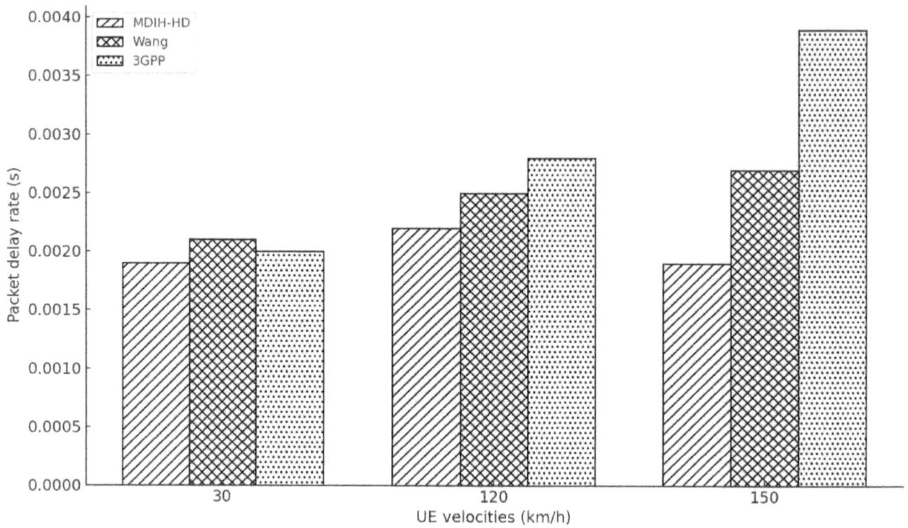

FIGURE 4.6 Comparison of packet delay rate using both applications in three handover algorithms.

TABLE 4.4
The percentage of MDIH-HD reduction in packet delay rate compared to Wang and 3GPP algorithms in both applications in one scenario

UE Velocity	Reduction Percentage Compared to Wang	Reduction Percentage Compared to 3GPP
30	8.2%	6.8%
120	7%	19%
150	23%	47.4%

The MDIH-HD algorithm has the lowest packet delay rate in the two applications compared to the other two algorithms. The reason behind this is due to the efficiency of the selection mechanism for the convenient target eNB which results in a reduction in handovers number and thus low failure link rate. Moreover, selecting the proper RSRP threshold value for triggering the handover also affects the reduction in the packet delay rate. Even more, using UE's table allows the UE to move directly between serving and target eNBs, and this will reduce the searching time for selecting the target eNB as well as will result in a lower data delay rate. Contrariwise, Wang algorithm obtained a good result at high UE velocity compared to 3GPP scheme due to the target eNB selection manner. The target eNB is selected depending on the smallest front angle and closest distance to the UE position. On the other hand, the result is worse than MDIH-HD because of searching for the candidate eNB list which is located ahead of the UE without the consideration of avoiding the eNBs which are far away from the current UE position. Even more, Wang algorithm depends on the UE direction for selecting the target eNB in all the cases while the MDIH-HD uses UE table for handover decision in routine routes.

4.2.4 Packet Loss Ratio

By increasing the UE velocity from 30 to 150 km/h as shown in Figure 4.7, the MDIH-HD algorithm shows an increase in the packet loss ratio in all its applications (CBR, VoIP, and both).

The CBR application has the highest packet loss ratio because of its packet size. The packet loss ratio increases from 0.0016 to 0.0066 s when the UE velocity increases from 30 to 150 km/h. With regard to VoIP application, the same figure shows that it has the lowest loss ratio because of the packet size and priority. The packet loss increases from 0.0003 to 0.0007 s when the UE velocity increases from 30 to 150 km/h. Moreover, when applying both applications in the same scenario, the same figure shows the medium result of packet loss ratio and that is because the packets size is divided between the two applications. The packet loss increases from 0.0014 to 0.0020 s when the UE velocity increases from 30 to 150 km/h. Regarding CBR application in the three handover decision algorithms when using different high UE velocities (30, 120, and 150 km/h), Figure 4.8 shows that there is an increment in the packet loss ratio in all handover decisions as the UE velocity increases, due to the effect of the velocity on the handover decision procedure.

The MDIH-HD increases the packet loss ratio from 0.0016 to 0.0066 s when the UE velocity increases from 30 to 150 km/h, and in Wang the packet loss ratio is increased from 0.0034 to 0.0091 s with the same change in UE velocity. Moreover, the packet loss ratio is increased from 0.0070 to 0.0144 s with the same change in UE

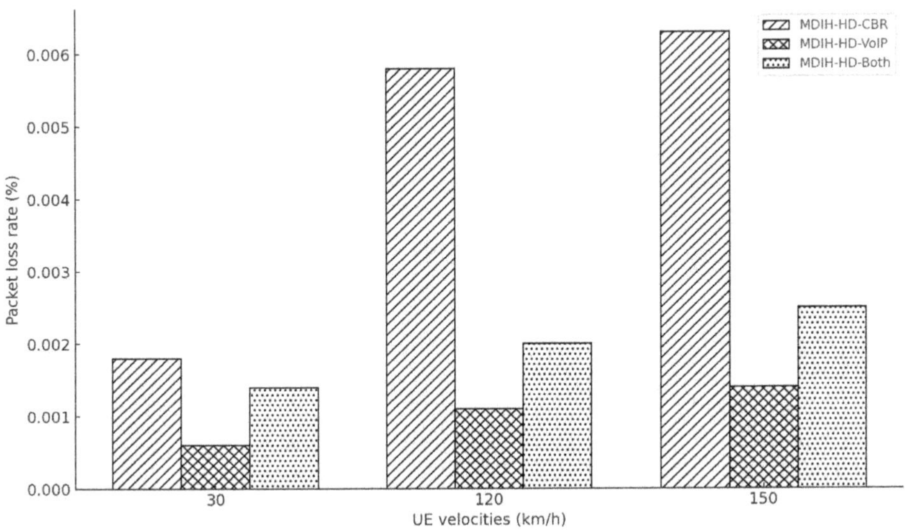

FIGURE 4.7 Comparison of packet loss ratio between applications in MDIH-HD algorithm.

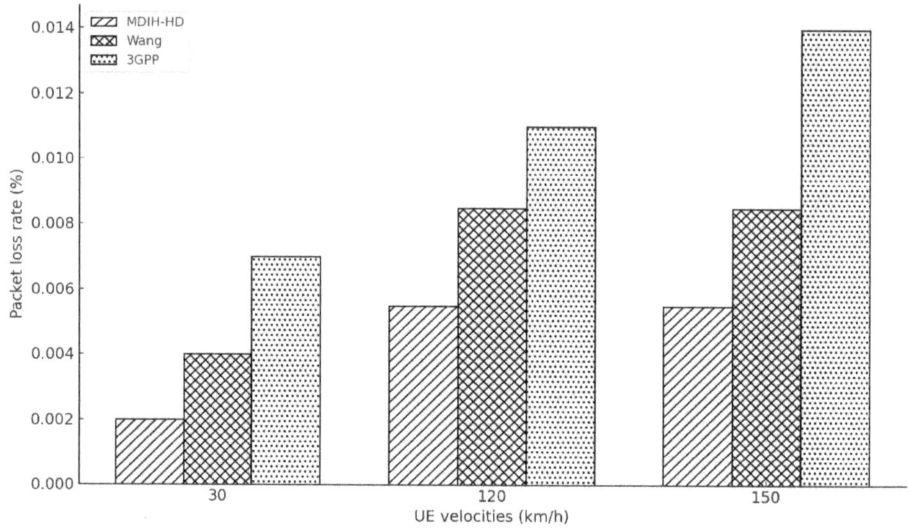

FIGURE 4.8 Comparison of packet loss ratio using CBR application in three handover algorithms.

TABLE 4.5
The percentage of MDIH-HD reduction in packet loss ratio compared to Wang and 3GPP algorithms in CBR application

UE Velocity	Reduction Percentage Compared to Wang	Reduction Percentage Compared to 3GPP
30	52%	76%
120	33%	47%
150	28%	54%

velocity. This increment in the packet loss ratio with the increase in the UE velocity is predictable. Thus, the handover in high velocities needs to be faster and more reliable in order to improve the QoS for UEs.

However, the proposed MDIH-HD algorithm reduces the packet loss ratio in comparison to Wang and 3GPP handover algorithms in the CBR application at every UE velocity as presented in Table 4.5.

Finally, concerning the packet loss ratio in the CBR application, it is reduced in the MDIH-HD algorithm by 37.6% compared to Wang algorithm, and by 47% compared to 3GPP handover algorithm for all different UE velocities (30, 120, and 150 km/h).

Regarding the VoIP application in the same scenario of UE velocities and as illustrated in Figure 4.9, the MDIH-HD increases the packet loss ratio from 0.0003 to 0.0007 s when the UE velocity increases from 30 to 150 km/h, while in Wang algorithm the packet loss ratio increases from 0.0008 to 0.0009 s with the same change in

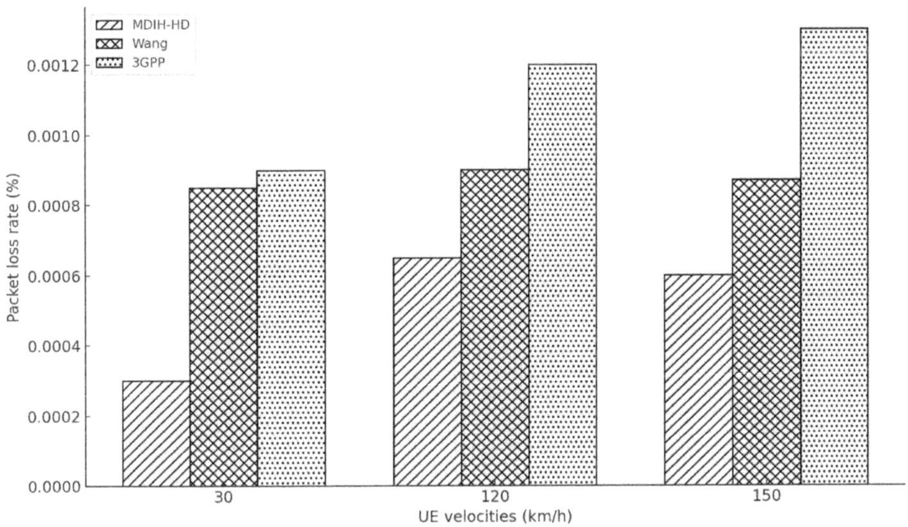

FIGURE 4.9 Comparison of packet loss ratio using VoIP application in three handover algorithms.

TABLE 4.6
The percentage of MDIH-HD reduction in packet loss ratio compared to Wang and 3GPP algorithms in VoIP application

UE Velocity	Reduction Percentage Compared to Wang	Reduction Percentage Compared to 3GPP
30	60%	64%
120	14%	40%
150	27%	49%

UE velocity. Moreover, the packet loss ratio increases from 0.0009 to 0.0012 s with the same change in UE velocity.

However, the proposed MDIH-HD algorithm reduces the packet loss ratio in comparison to Wang and 3GPP handover algorithms of the VoIP application at every UE velocity as presented in Table 4.6.

Finally, concerning the packet loss ratio in the VoIP application, it is reduced in the MDIH-HD algorithm by 33.6% compared to Wang algorithm, and by 51% compared to 3GPP handover algorithm for all different UE velocities (30, 120, and 150 km/h).

In addition, when both applications were applied in the same scenario using the three handover decision algorithms in different high UE velocities (30, 120 and 150 km/h), the results are shown in Figure 4.10.

Based on Figure 4.10, the MDIH-HD increases the packet loss ratio from 0.0014 to 0.0020 s when the UE velocity increases from 30 to 150 km/h, while in Wang algorithm the packet loss ratio increases from 0.0016 to 0.0043 s with the same change

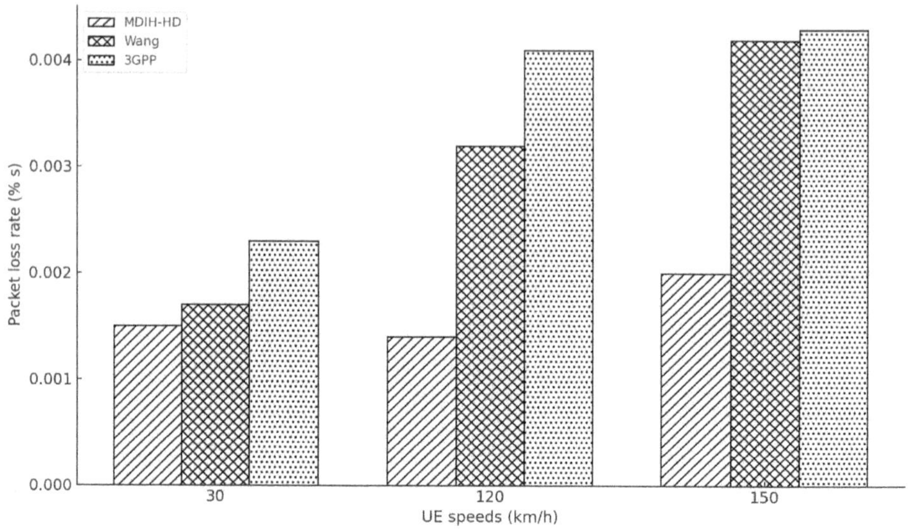

FIGURE 4.10 Comparison of packet loss ratio using both applications in three handover algorithms.

TABLE 4.7
The percentage of MDIH-HD reduction in packet loss ratio compared to Wang and 3GPP algorithms in both applications in one scenario

UE Velocity	Reduction Percentage Compared to Wang	Reduction Percentage Compared to 3GPP
30	13.2%	40%
120	55%	64%
150	52%	52%

in UE velocity. Moreover, the packet loss ratio increases in the 3GPP handover algorithm from 0.0024 to 0.0043 s with the same change in UE velocity.

However, the proposed MDIH-HD algorithm reduces the packet loss ratio in comparison to Wang and 3GPP handover algorithms in both applications at every UE velocity as demonstrated in Table 4.7.

Finally, concerning the packet loss ratio in both applications, it is reduced in the MDIH-HD algorithm by 40% compared to Wang algorithm and by 52% compared to 3GPP handover algorithm for all different UE velocities (30, 120, and 150 km/h).

The MDIH-HD algorithm shows the lowest packet loss ratio in all applications compared to the other two algorithms. The reason behind this is due to the efficient use of UE table in handover decision. The use of UE table in the routine routes assists the UE in selecting the target eNB directly and saving the time consumed in searching for and selecting the target eNB. Even more, selecting the target eNB by using the moving direction procedure also reduces the packet loss ratio through avoiding the neighbor eNBs which exist out of the range of the UE direction and far away from UE position.

4.2.5 System Throughput

When the UE velocity increases from 30 to 150 km/h, as shown in Figure 4.11, the MDIH-HD algorithm shows a decrease in the system throughput in all the applications (CBR, VoIP, and both).

The CBR application has the highest throughput because of the packet size. The throughput decreases from 2939.94 to 2925.28 kbps when the UE velocity increases from 30 to 150 km/h. Regarding the VoIP application, the same figure shows that it has the lowest throughput because of the packet size and priority. The throughput decreases from 527.45 to 527.26 kbps when the UE velocity increases from 30 to 150 km/h. Moreover, when applying both applications in the same scenario, the same figure shows the medium result of throughput and that is because the packet sizes are divided between the two applications. The throughput decreases from 1752.40 to 1750.18 kbps when the UE velocity increases from 30 to 150 km/h.

Indeed, and as shown in Figure 4.12, when there are different mobility velocities (30, 120, and 150 km/h) in CBR application, the throughput of MDIH-HD decreases from 2939.94 to 2925.28 kbps when the UE velocity increases from 30 to 150 km/h. In Wang algorithm, the throughput decreases from 2934.64 to 2917.76 kbps whereas

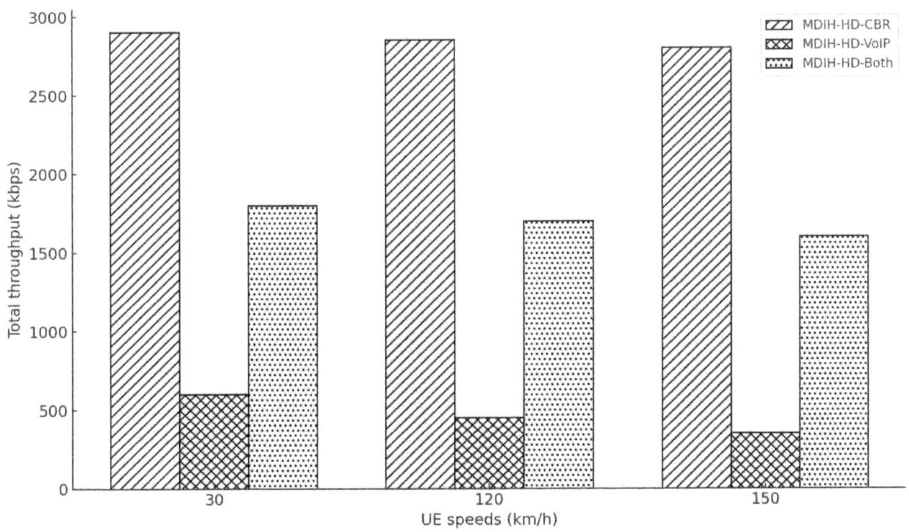

FIGURE 4.11 Comparison of system throughput between applications in MDIH-HD algorithm.

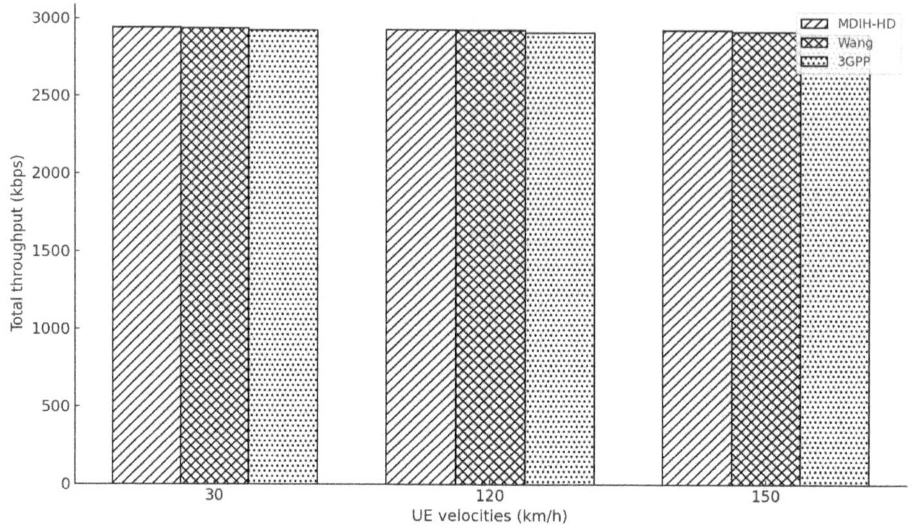

FIGURE 4.12 Comparison of system throughput using CBR application in three handover algorithms.

TABLE 4.8
The percentage of MDIH-HD improvement in throughput compared to Wang and 3GPP algorithms in CBR application

UE Velocity	Reduction Percentage Compared to Wang	Reduction Percentage Compared to 3GPP
30	0.18%	0.54%
120	0.3%	0.54%
150	0.25%	0.79%

it decreases from 2923.96 to 2902.08 kbps in 3GPP with the same change in UE velocity.

Finally, concerning the system throughput in both applications, in average it is increased in the MDIH-HD algorithm by 0.24% compared to Wang algorithm and by 0.62% compared to 3GPP handover algorithm for all different UE velocities (30, 120, and 150 km/h) in CBR application as demonstrated in Table 4.8.

Regarding the VoIP application under the same UE velocity scenarios illustrated in Figure 4.13, the throughput in the MDIH-HD algorithm decreases from 527.45 kbps to 527.26 kbps when the UE velocity increases from 30 km/h to 150 km/h, whereas in Wang algorithm the throughput decreases from 527.20 to 527.12 kbps while maintaining the same UE velocity. Moreover, in the 3GPP algorithm, the throughput is decreased from 527.01 kbps to 526.97 kbps also with the same change in the UE velocity.

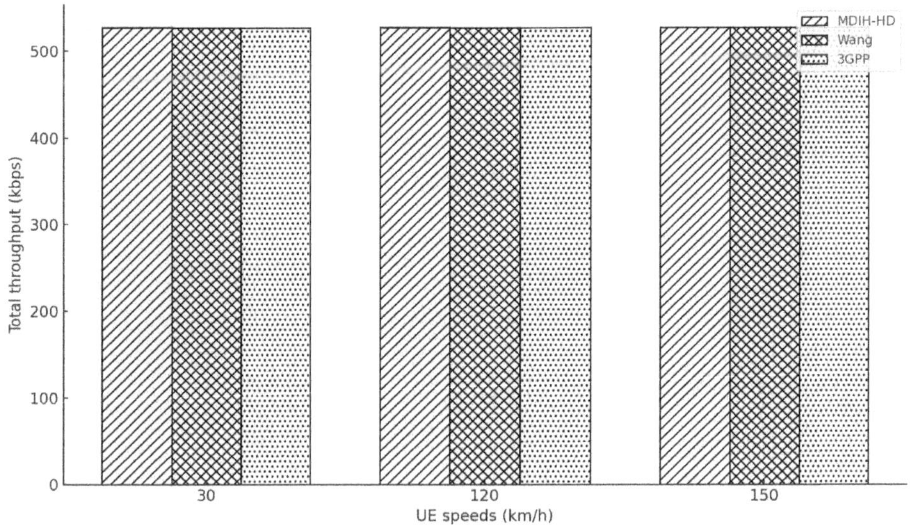

FIGURE 4.13 Comparison of the system throughput using VoIP application in three handover algorithms.

TABLE 4.9
The percentage of MDIH-HD improvement in throughput compared to Wang and 3GPP algorithms in VoIP application

UE Velocity	Reduction Percentage Compared to Wang	Reduction Percentage Compared to 3GPP
30	0.05%	0.08%
120	0.11%	0.24%
150	0.25%	0.53%

However, the proposed MDIH-HD algorithm improved the throughput in comparison to Wang and 3GPP handover algorithms in VoIP application at every UE velocity as demonstrated in Table 4.9.

Finally, concerning the system throughput in both applications, in average it is increased in the MDIH-HD algorithm by 0.13% compared to Wang algorithm and by 0.28% compared to 3GPP handover algorithm for all different UE velocities (30, 120, and 150 km/h) in VoIP application as demonstrated in Table 4.9.

Regarding both applications under the same UE velocity scenario illustrated in Figure 4.14, the improvement in the system throughput in the MDIH-HD algorithm is slightly higher than in Wang when the UE's velocity is 30 km/h. However, the improvements are considerable when UE velocity is either 120 or 150 km/h. On the contrary, the MDIH-HD algorithm shows improvement in the system throughput at all UE velocities compared to 3GPP handover algorithm. The MDIH-HD decreases

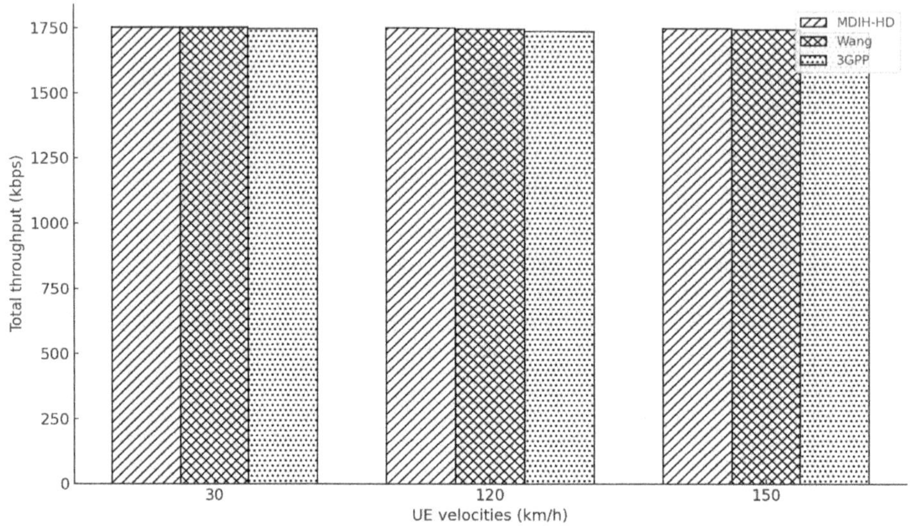

FIGURE 4.14 Comparison of system throughput using both applications in three handover algorithms.

TABLE 4.10
The percentage of MDIH-HD improvement in the system throughput compared to Wang and 3GPP algorithms in both applications

UE Velocity	Reduction Percentage Compared to Wang	Reduction Percentage Compared to 3GPP
30	0.02%	0.23%
120	0.23%	0.66%
150	0.18%	0.60%

the throughput from 1752.40 to 1750.18 kbps when the UE velocity increases from 30 to 150 km/h, and in Wang algorithm the throughput is decreased from 1752.00 to 1747.04 kbps, with the same UE velocity. Moreover, in 3GPP standard handover algorithm, the throughput is decreased from 1748.23 to 1739.63 kbps, with the same change in the UE velocity.

Finally, concerning the system throughput in both applications, in average it is increased in the MDIH-HD algorithm by 0.14% compared to Wang algorithm and by 0.49% compared to 3GPP handover algorithm for all different UE velocities (30, 120, and 150 km/h) in both applications as demonstrated in Table 4.10.

Consequently, as the results show, it can be noticed that by increasing the UE velocity, all handover decision procedures show degradation in the throughput. This is because of the high UE mobility that is needed to execute handover in a short period of time which leads to a higher possibility of a handover failure.

4.3 SUMMARY

The performance of the MDIH-HD algorithm is investigated and validated in this chapter. Additionally, parameters setting that is related to the MDIH-HD algorithm is discussed. The MDIH-HD algorithm uses S parameter (distance between two UE positions) in tracking UE positions to predict the future UE direction while uses distance and angle parameters to select the target eNB. Moreover, it uses the UE table for selecting the target eNB directly in the routine routes. In addition, two applications (CBR and VoIP) are used to investigate the MDIH-HD performance in different high-mobility velocities. The MDIH-HD algorithm showed the best performance in terms of the average number of handovers, signaling measurements rate, packet delay rate, packet loss ratio, and system throughput compared to present handover decision algorithms. This improvement in performance of the MDIH-HD algorithm is related to storing UE paths in the UE tables which are reflected in reducing the target eNB selection time. Moreover, in the case when the UE path is not available in the UE table, the moving direction prediction procedure is executed to reduce the searching time for target eNB through avoiding the neighbor eNBs that are not in front of UE trajectory. In addition, the MDIH-HD algorithm obviously works better than the 3GPP and Wang et al. (2014a) handover decision algorithms.

5 Moving Direction Distance-Vertical Handover Decision Algorithm

5.1 INTRODUCTION

In this chapter, a new MDD-VHD algorithm is proposed based on UE trajectory, UE velocity, and cell location to improve the vertical handover decision in LTE-A. The performance of the proposed HDD-VHD is presented and discussed. The simulation uses constant UE velocity, which is 25 km/h, two eNBs, incremental numbers of UEs configured as 10, 20, 40, and 80, and incremental numbers of HeNBs configured as 20, 40, 60, 80, and 160 in two traffic applications (CBR and VoIP). In this chapter, the performance of the HDD-VHD is evaluated using five performance metrics for each UE group and each HeNB density in the three distribution scenarios. The performance metrics are the average handover number, average signaling measurement number, packet delay, packet loss, and throughput. Moreover, the performance of the HDD-VHD algorithm is compared with the handover decision algorithm in Deswal and Singhrova (2016). In Deswal and Singhrova (2016), the authors increased the RSS threshold value for the handover procedure when it takes place from HeNB to eNB. On the contrary, they decreased the RSS threshold value for the handover procedure when it takes place from eNB to HeNB. Finally, the results of the proposed algorithms are presented and discussed, and the conclusion remarks are given.

5.2 PERFORMANCE OF MDD-VHD ALGORITHM

In this simulation experiment for evaluating the proposed MDD-VHD algorithm, comparisons and evaluations of the performance are made using the following performance metrics: average of handovers number, average signaling measurements rate, packet delay rate, packet loss ratio, and throughput. Moreover, the simulation work was carried out by integrating the LTE-Sim simulators, and each simulation was carried out ten times in order to obtain the average results values for each UEs number and for each scenario of HeNBs distribution in each application. The simulation subresults and analysis based on these metrics will be shown and analyzed in the following subsections. The whole results of MDD-VHD scenarios and reference in both applications are expounded in further detail in Appendix B and Appendix C.

DOI: 10.1201/9781003648130-5

5.2.1 Average Handover Number

Regarding the three scenarios of HeNBs distribution (close, middle, and at the edge), the HeNBs are distributed to groups of 20, 40, 60, and 80 in each scenario. Figure 5.1 shows the average number of handovers for the MDD-VHD scheme in each scenario when the UEs number is configured as 20. As illustrated in Figure 5.1, the density of HeNBs in the close area increases the average of handovers number, while it has the lowest average when the HeNBs distribution is at the edge. This is because the UEs start moving from the location of eNB tower. In addition, the average of handovers number increases as the number of HeNBs in all the distribution scenarios increases.

Furthermore, the results of average handovers number for the proposed HDD-VHD and the Deswal handover algorithms were discussed in terms of HeNBs that are distributed to groups of 20, 40, 60, and 80 per each eNB cell, and two groups of UEs (10 and 20) as presented in Figure 5.2. Based on each result, it is evident that by increasing the HeNBs number, both algorithms show an increase in the average handovers number because UEs make additional handovers with respect to their movements in each UE group.

The average handovers number of the MDD-VHD algorithm increases from 0.03 to 0.161 when HeNBs number increases from 20 to 80 in the case of using 10 UEs. Moreover, the average of handovers number increases from 0.061 to 0.338 with the same change in HeNBs number and UEs number in Deswal algorithm. In addition, the same figure shows the results of average handovers number when using 20 UEs and the same variation in HeNBs number. The average of handovers number is higher in both algorithms than the average of handovers number when the number of UEs is 10 and this is due to the increase in UEs density. The average handovers number

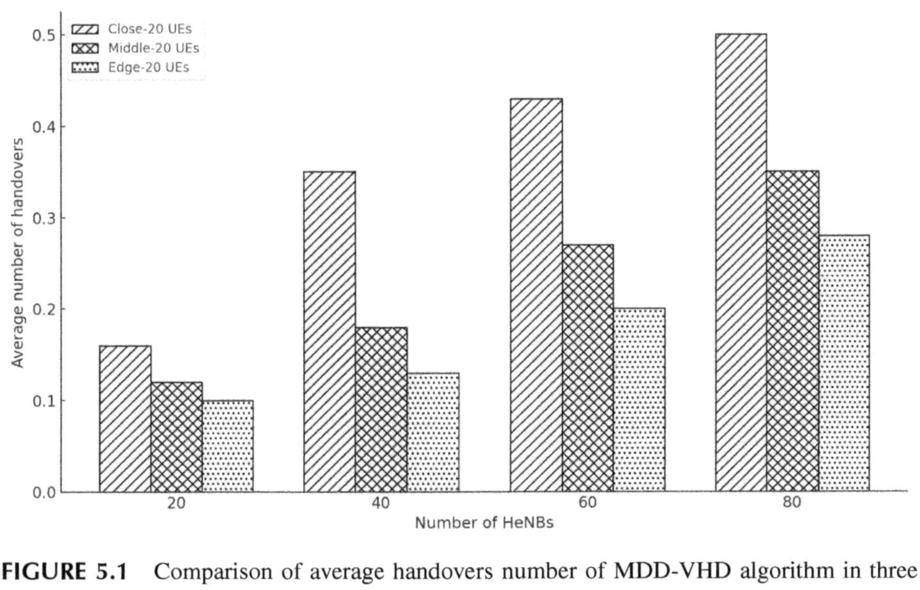

FIGURE 5.1 Comparison of average handovers number of MDD-VHD algorithm in three scenarios.

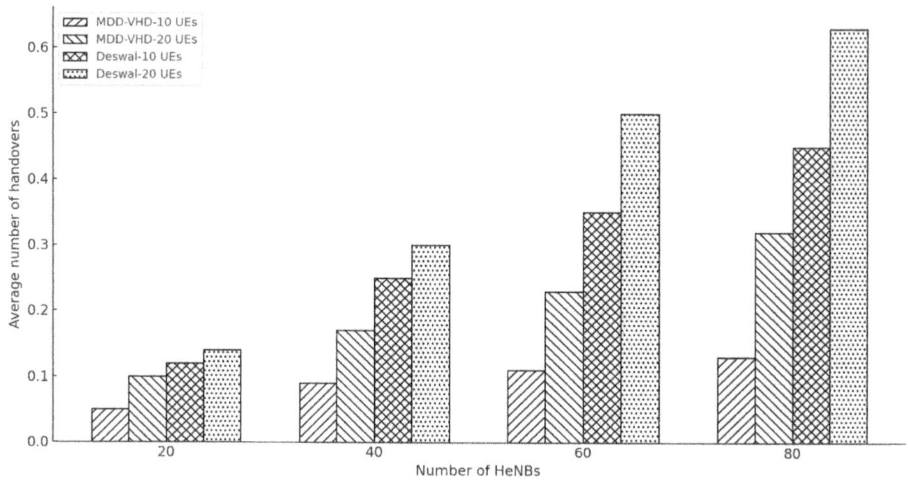

FIGURE 5.2 Comparison of average handovers number of both algorithms.

of the MDD-VHD algorithm increases from 0.092 to 0.407 when HeNBs number increases from 20 to 80. Additionally, the average of handovers number increases from 0.130 to 0.676 with the same change in HeNBs number in Deswal algorithm.

The results emphasize that the best performance was achieved by the MDD- VHD algorithm in all the distributions of HeNBs and all densities. This is because the handover decision in the proposed algorithm does not connect to HeNBs that are close to the UE, but only to those located in front of or approximately ahead of current UE position in order to avoid the redundant handover. The UE only nominates the HeNB whose tower location is less than |25±| and the distance between the UE and the candidate HeNB is less than or equal to 28 m. Moreover, using a polynomial function to know the UE trajectory also helps the UE to select the best target HeNB. On the contrary, in the case of Deswal handover decision the handover procedure is triggered when the RSRP between the UE and its neighbor HeNBs is higher than the RSRP between the UE and its serving cell without any consideration of how long the target HeNB will serve the UE and its usefulness to do handover or not. Therefore, the average handovers number is higher than it is in MDD-VHD.

However, the proposed MDD-VHD algorithm reduces the average number of handovers compared to the Deswal handover for each UE group and in each HeNBs density as presented in Table 5.1.

Finally, regarding the total average of handovers number for each UE group, it is reduced in the MDD-VHD algorithm by 48% compared to Deswal handover algorithm for all various HeNBs distributions and densities when the number of UEs is 10. Moreover, the MDD-VHD algorithm reduces the total average of handovers number by 43.2% compared to Deswal handover algorithm for all various HeNBs distributions and densities when the number of UEs is 20.

TABLE 5.1
The percentage of MDD-VHD reduction in average of handovers number compared to Deswal

UE Group	HeNB	Reduction Percentage
10	20	50.0%
10	40	33.0%
10	60	60.0%
10	80	52.0%
20	20	29.0%
20	40	54.0%
20	60	50.0%
20	80	40.0%

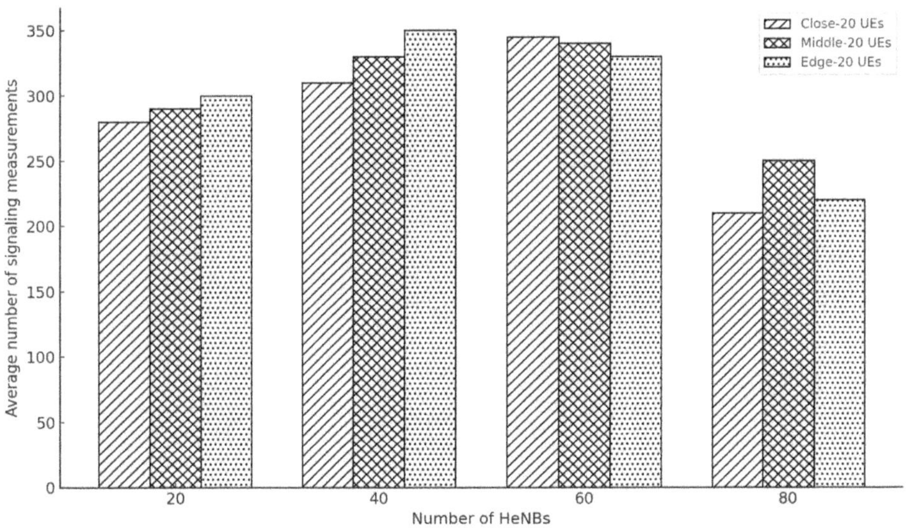

FIGURE 5.3 Comparison of average number of signaling measurements of MDD-VHD in three scenarios.

5.2.2 Average Signaling Measurements Number

With regard to the three scenarios of HeNBs distribution with different HeNB densities and when the UEs number is configured as 20, Figure 5.3 shows that the average rate of signaling measurements of MDD-VHD is not significantly affected by the density of HeNBs. The UE performs singling measurement with the cells which are in front of and close to its position while ignoring the rest. Moreover, the variation in the average number of signaling measurements, as in Figure 5.3, is related to HeNB positions and UEs direction regardless of the HeNBs density.

By comparing the average of signaling measurements between the MDD-VHD and existing work when increasing HeNBs density from 20 to 80 cells in both UE groups as shown in Figure 5.4, it is found that the MDD-VHD algorithm shows a sharp decrease in the average signaling measurements number. It is noticed that the average number of signaling measurements in the Deswal handover scheme increases regularly when the HeNBs number and UEs number increase. Thus, the Deswal handover procedure proved that the duration of signaling measurement is fixed on the basis of the cells density. Moreover, the Deswal handover decision applies the signaling measurement control on all UE neighbor cells in each measured period through comparing the RSRP of the UE with its serving cell, on one hand, and with UE neighbor cells, on the other hand.

Nevertheless, our proposed scheme significantly reduces the number of signaling measurements compared to Deswal algorithm for each UE group. This is due to the use of UE trajectory prediction on one hand, and the use of distance parameter between HeNB tower location and the current position of UE in HeNBs cells nomination on the other hand, which is set to 28 m. Therefore, the UE performs signaling measurement only with the HeNBs which are in the candidate list in order to select the target HeNB and avoid the remaining neighbor cells. In addition, the UE performs signaling measurement with the candidate HeNB list for each *S* distance between the current position of UE and the previous UE position, and this is another reason for the reduction in the signaling measurements average number.

The average signaling measurements number of MDD-VHD increases from 111.4 to 142.9 when HeNBs density increases from 20 to 80 in the case of using 10 UEs. In Deswal algorithm, the average signaling measurements number changes in

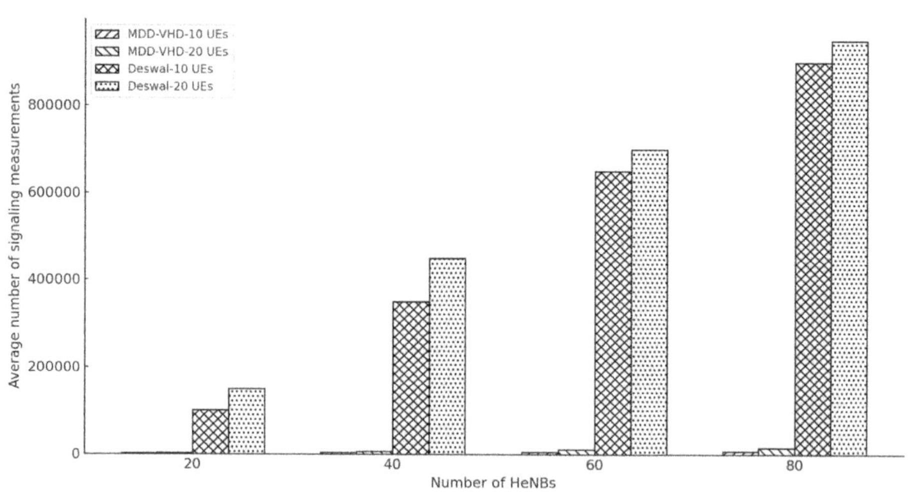

FIGURE 5.4 Comparison of average signaling measurements number in two handover algorithms.

a regular manner as it increases from 118,342 to 473,369 when the HeNBs density increases from 20 to 80 with the same number of UEs. In addition, when the number of UEs is 20 and the same variation in HeNBs number is used, the average of signaling measurements number is increased compared to the average of signaling measurements number when the UEs number is 10, and this is related to the increase in UEs density. The average signaling measurements number of MDD-VHD algorithm changes from 158 to 325 when the HeNBs number changes from 20 to 80. In addition, the average of handovers number increases from 236,684 to 946,738 with the same change in HeNBs number in the Deswal algorithm.

However, the total average of signaling measurements rate in MDD-VHD is reduced by more than 99% compared to Deswal algorithm in each UE group, and that means a reduction in UE's power consumption.

5.2.3 Packet Delay Rate

Regarding the three scenarios of HeNBs distribution (close, middle, and at the edge), the HeNBs are distributed to groups of 20, 40, 60, and 80 in each scenario and the UEs number is configured as 20. Figure 5.5 and Figure 5.6 present the simulation results of the MDD-VHD algorithm of CBR and VoIP applications in terms of packet delay rate.

The two figures show that with an increase in the number of HeNBs, all distribution scenarios show an increment in packet delay rate. Moreover, both figures show that the lowest packet delay rate occurs when the density of HeNBs is concentrated in the middle and the highest one is when the HeNBs are distributed at the edge. This is because UE starts moving from the location of eNB tower and the density of

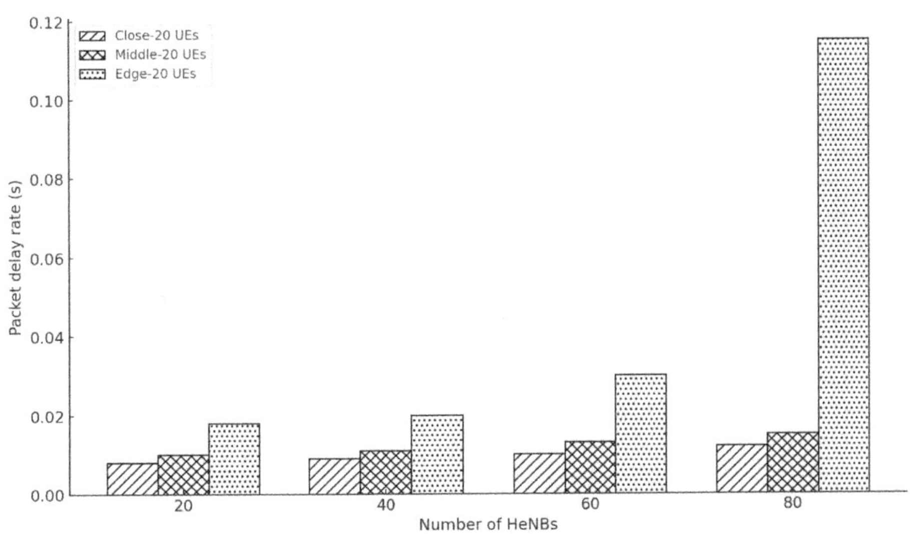

FIGURE 5.5 Comparison of packet delay rate of MDD-VHD algorithm in three scenarios by using CBR application.

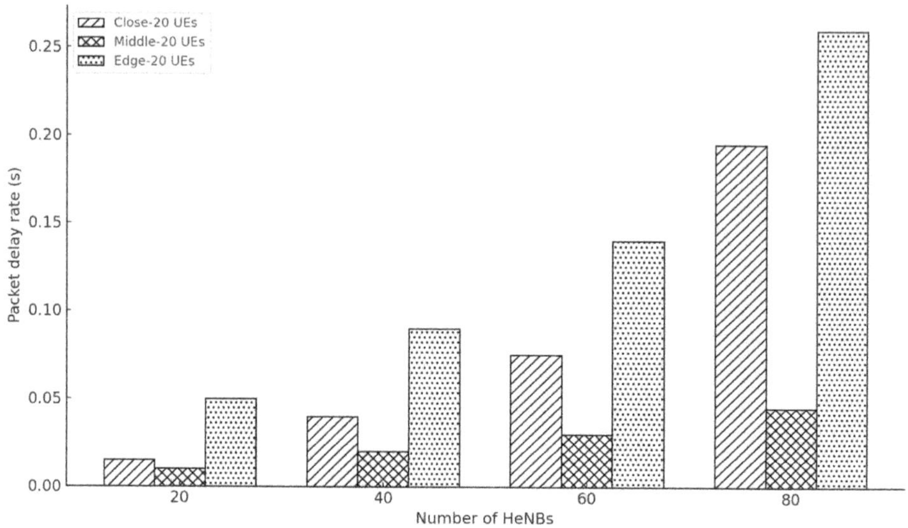

FIGURE 5.6 Comparison of packet delay rate of MDD-VHD algorithm in three scenarios by using VoIP application.

HeNBs, when they are distributed close to eNB tower, affects the packet delay rate by increasing the handovers number and interference. When HeNBs are distributed at the edge of the eNB tower, the packet delay is increased because of the weakness of RSS between the UE and its serving eNB. In addition, it is observed from the two figures that the packet delay rate of the VoIP application is more than that of the CBR application. This is because of the *Maximum–delay* parameter which is configured as 0.1 s in both applications and the VoIP application is affected by the packet delay rate more than the CBR because of the packet size and the priority of real-time application.

Regarding the comparison between the MDD-VHD and existing work in terms of packet delay rate, two handover decision algorithms were evaluated using sets of different HeNBs density (20, 40, 60, and 80) for each group of UEs (10 and 20) in CBR application. Figure 5.7 shows that both handover decision algorithms increment the packet delay rate when the HeNBs number increases because of the effect of HeNBs number on the number of handover procedures. Thus, the handover inside a large number of HeNBs will be frequent and the delay in selecting the appropriate target cell during the searching time increases the packet delay rate.

The MDD-VHD algorithm of CBR application increases the packet delay rate from 0.0021 to 0.0053 s when the HeNBs number increases from 20 to 80 and in the case of using 10 UEs, while in Deswal algorithm the packet delay rate increases from 0.0037 to 0.5309 s with the same change in HeNBs number and the same UEs number. Moreover, in the case of using 20 UEs with the same change in HENBs number, the packet delay rate shows a variation in the results compared to the packet delay rate when the UEs number is 10 in each algorithm. The packet delay rate in the MDD-VHD algorithm increases from 0.0022 to 0.0129 s when the HeNBs number

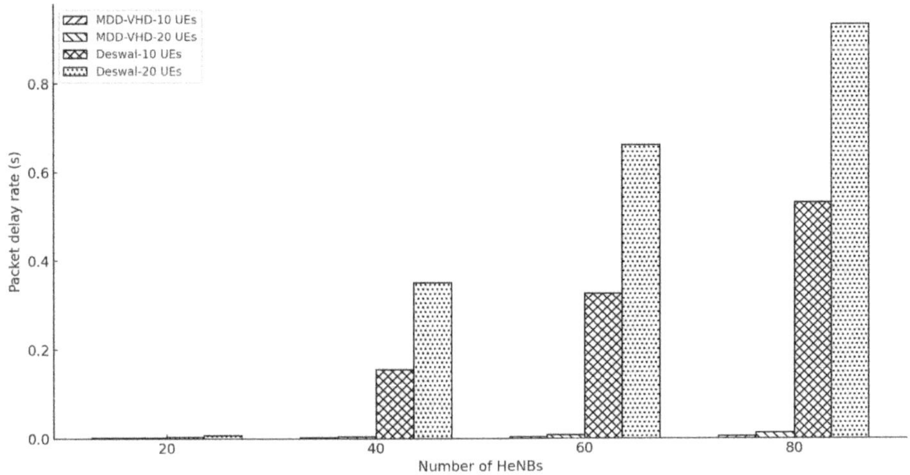

FIGURE 5.7 Comparison of packet delay rate using CBR application in two handover algorithms.

TABLE 5.2
The percentage of MDD-VHD reduction in packet delay rate compared to Deswal using CBR application

UE Group	HeNB	Reduction Percentage
10	20	42.0%
10	40	97.0%
10	60	98.0%
10	80	99.0%
20	20	69.0%
20	40	98.0%
20	60	99.0%
20	80	98.0%

increases from 20 to 80. In addition, the packet delay rate increases from 0.0074 to 0.9324 s with the same change in HeNBs number in Deswal algorithm.

However, the proposed MDD-VHD algorithm reduces the packet delay rate compared to the Deswal handover for each of the two UE groups and in each HeNBs density in CBR application as illustrated in Table 5.2.

Table 5.2 shows that the efficiency of the MDD-VHD algorithm is improved with an increase in the number of HeNBs compared to the Deswal algorithm. In general, the MDD-VHD algorithm in CBR application reduces the packet delay rate by 87.5% compared to Deswal handover algorithm in all scenarios.

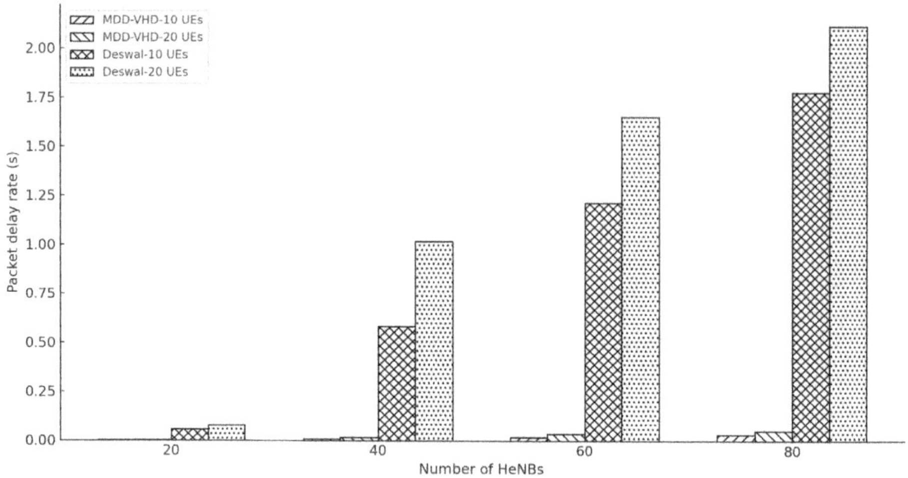

FIGURE 5.8 Comparison of packet delay rate using VoIP application in two handover algorithms.

With regard to the VoIP application, the two handover decision algorithms were evaluated using a different number of HeNBs (20, 40, 60, and 80) and a specific number of UEs (10 and 20). Figure 5.8 shows that both handover decision algorithms increase the packet delay rate as the HeNBs number increases. Furthermore, the packet loss rate using VoIP application is higher compared to CBR application.

The MDD-VHD algorithm in VoIP application increases the packet delay rate from 0.0014 to 0.0313 s when the HeNBs number increases from 20 to 80 and the number of UEs is 10, while in Deswal handover algorithm the packet delay rate increases from 0.0601 to 1.7760 s with the same change in HeNBs number and when using ten UEs. Moreover, in the case of using 20 UEs and the same change in HeNBs number, the packet delay rate shows the variation in the results in comparison to the packet delay rate when using ten UEs in both algorithms. The packet delay rate in the MDD-VHD algorithm increases from 0.0022 to 0.0511 s when the HeNBs number increases from 20 to 80. In addition, the packet delay rate increases from 0.07952 to 2.11548 s with the same change in HeNBs number in the Deswal algorithm.

However, the proposed MDD-VHD algorithm reduces the packet delay rate compared to the Deswal handover for each UE group and in each HeNBs density in VoIP application as illustrated in Table 5.3.

It is observed from Table 5.3 that the efficiency of the MDD-VHD algorithm has improved with an increase in the number of HeNBs compared to the Deswal algorithm. In general, the MDD-VHD algorithm in VoIP application reduces the packet delay rate by 98.5% compared to Deswal handover algorithm in all scenarios. However, the MDD-VHD algorithm shows the lowest packet delay rate compared to Deswal algorithm in both applications in each UEs group and in each HeNBs group. The reason as to why the MDD-VHD has a lower packet delay rate is attributed to the efficient selection mechanism of the convenient target HeNB, which leads to a

TABLE 5.3
The percentage of MDD-VHD reduction in packet delay rate compared to Deswal in VoIP application

UE Group	HeNB	Reduction Percentage
10	20	97.0%
10	40	99.0%
10	60	99.0%
10	80	98.0%
20	20	97.0%
20	40	99.0%
20	60	98.0%
20	80	97.0%

reduction in handovers number and thus lower connection and pre-connection phases. Avoiding neighbor HeNBs that are useless from handovers is the main goal of using the UE trajectory together with the distancing technique since the HeNB coverage area is very small compared to eNB. Therefore, the UE in VDD-VHD algorithm connects as long as possible with the convenient target HeNB which met the UE direction. On the contrary, the Deswal handover connects with each neighbor HeNB that has a stronger RSRP than the UE serving cell, and this will increase the number of frequent handovers, searching time, and pre-connection and connection phases. This results in raising the number of frequent handovers and in this case the packet delay and RLF will be increased.

5.2.4 Packet Loss Ratio

Regarding the three scenarios of HeNBs distribution (close, middle, and at the edge), the HeNBs are distributed to groups of 20, 40, 60, and 80 in each scenario and the UEs number is configured as 20. Figure 5.9 and Figure 5.10 present the simulation results of the MDD-VHD algorithm using both CBR and VoIP applications in terms of packet loss ratio. It is observed from the two figures that the HeNBs number increases the packet loss ratio in all distribution scenarios.

Moreover, both figures show that the lowest packet delay rate is when the density of HeNBs is in the middle, while the highest rate is when the HeNB density is at the edge. This is because the UE starts moving from the eNB tower location and the density of HeNBs when concentrated close to eNB tower affects the packet loss by increasing both the handovers number and interference.

Furthermore, when HeNBs are distributed at the edge, the packet loss increases because of the weakness of RSS between the UE and its serving eNB while UE executes hand-out handover and the density of HeNBs is less than their density when distributed in the middle. In addition, Figure 5.9 and Figure 5.10 indicate that the

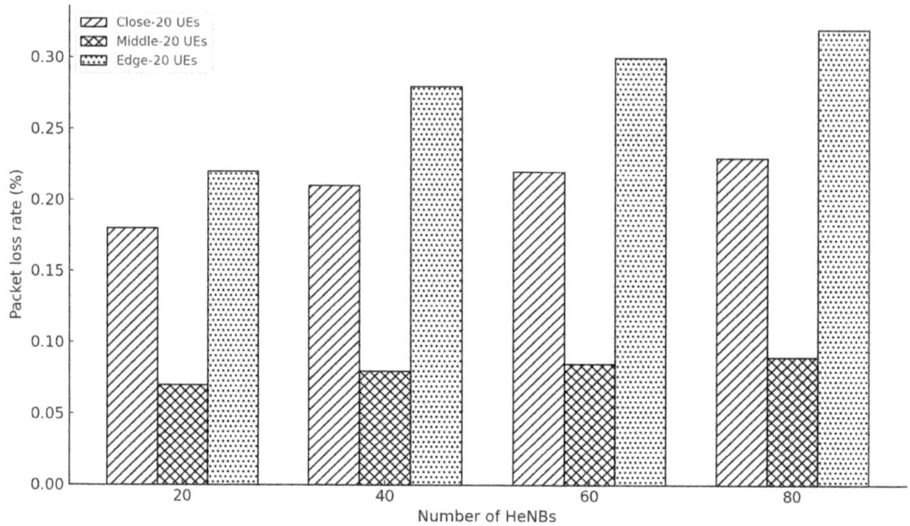

FIGURE 5.9 Comparison of packet loss ratio of MDD-VHD algorithm in three scenarios using CBR application.

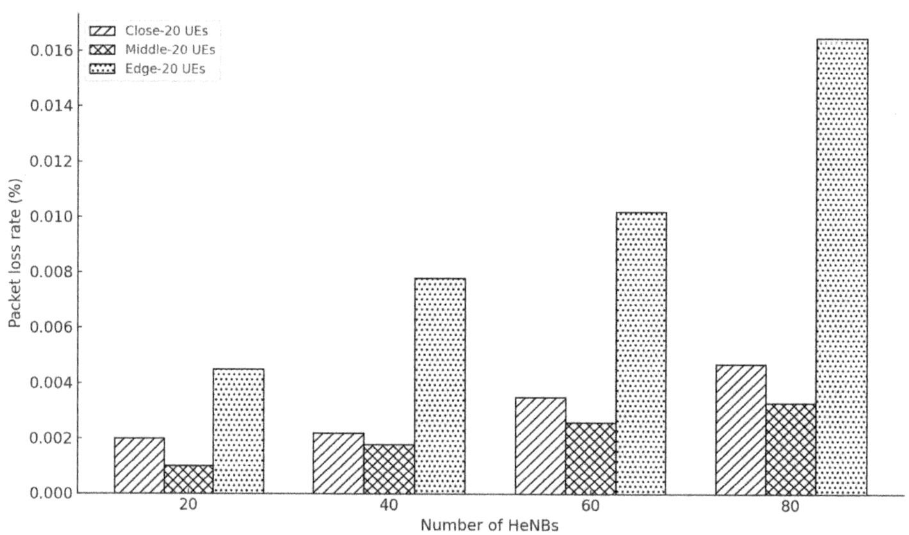

FIGURE 5.10 Comparison of packet loss ratio of MDD-VHD algorithm in three scenarios using VoIP application.

packet loss ratio using the CBR application is higher than when using the VoIP application. This is because of the difference in the packet size and the priority of real-time application. The packet size of CBR is 512 bytes with a rate of 8 packets/second while the VoIP packet size is 32 bytes per 20 ms time interval.

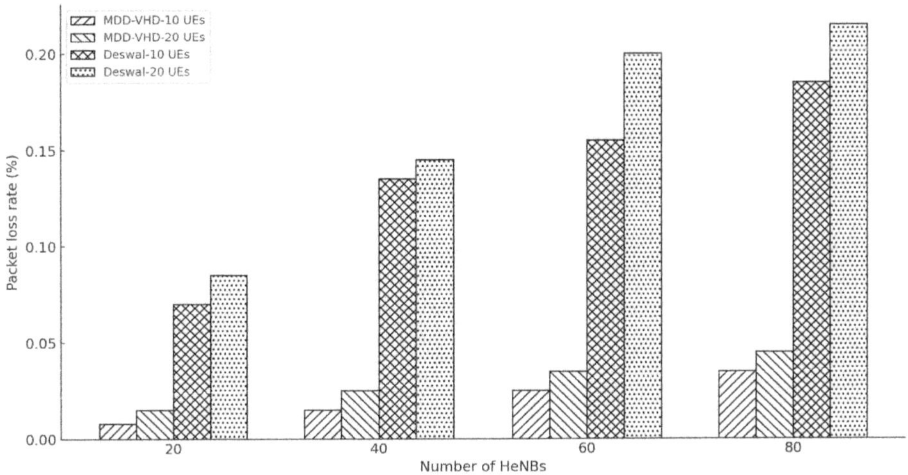

FIGURE 5.11 Comparison of packet loss ratio using CBR application in two handover algorithms.

By increasing the HeNBs density from 20 to 80 cells for each UE group (10 and 20) using CBR application, both algorithms show an increase in the packet loss ratio as explained in Figure 5.11. This increment of packet loss ratio is observed as long as the HeNBs number is predictable due to the frequent handovers between cells, especially among HeNBs, which cover a small area and each handover process has a probability of radio link failure. The MDD-VHD increases the packet loss ratio from 0.0045 to 0.0526 s when the HeNBs number increases from 20 to 80 and in the case of using ten UEs, while in Deswal algorithm the packet loss ratio increases from 0.0439 to 0.1998 s with the same change in HeNBs number and the same UEs number. In addition, in the case of using 20 UEs, the MDD-VHD algorithm increases the packet loss ratio from 0.0069 to 0.0585 s when the HeNBs number increases from 20 to 80, and in Deswal algorithm the packet loss ratio is increased from 0.0748 to 0.2187 s with the same change in HeNBs number using the same application.

However, the proposed MDD-VHD algorithm reduces the packet loss ratio in comparison to the Deswal handover for each of the two UE groups and in each density of HeNBs in CBR application as illustrated in Table 5.4.

As observed from Table 5.4, the MDD-VHD algorithm reduces the packet loss ratio compared to the Deswal algorithm. In general, the MDD-VHD algorithm in CBR application reduces the packet loss ratio by 82.6% compared to Deswal handover algorithm in all scenarios.

With regard to the VoIP application, the two handover decision algorithms were evaluated using a different number of HeNBs (20, 40, 60, and 80) and a specific number of UEs (10 and 20). Figure 5.15 shows that both handover decision algorithms increase the packet loss ratio as the HeNBs number increases. Furthermore, by comparing both applications in Figure 5.11 and Figure 5.12, it is observed that the packet

TABLE 5.4
The percentage of MDD-VHD reduction in packet loss ratio compared to Deswal in CBR application

UE Group	HeNB	Reduction Percentage
10	20	89.0%
10	40	88.0%
10	60	84.0%
10	80	73.0%
20	20	90.0%
20	40	83.0%
20	60	81.0%
20	80	73.0%

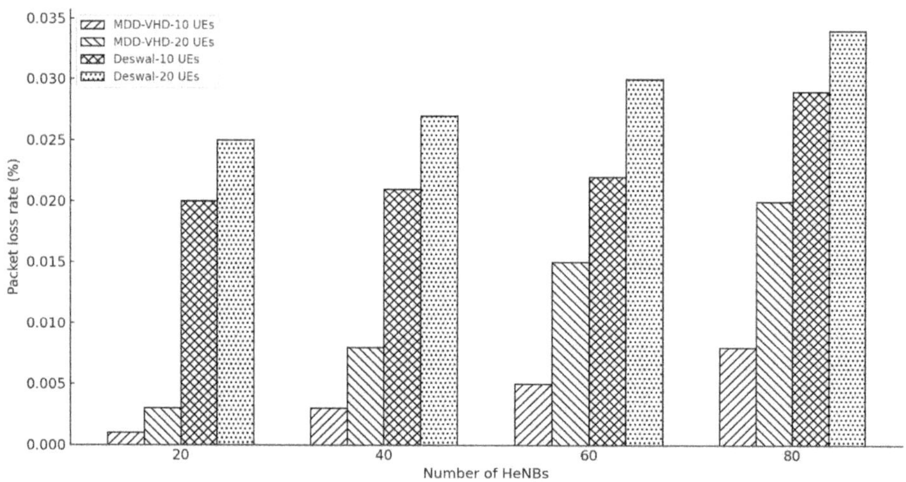

FIGURE 5.12 Comparison of packet loss ratio using VoIP application in two handover algorithms.

loss ratio using the CBR application is higher than the packet loss ratio using VoIP application.

The MDD-VHD algorithm in VoIP application increases the packet loss ratio from 0.00034 to 0.00582 s when the HeNBs number increases from 20 to 80 and the number of UEs is 10, while in Deswal handover algorithm the packet loss ratio increases from 0.00806 to 0.0248 s with the same change in HeNBs number and when using 10 UEs. Moreover, in the case of using 20 UEs and the same change in HeNBs number, the packet loss ratio shows the variation in the results in comparison to the packet loss ratio when using ten UEs in both algorithms because of the effect of UEs. The packet loss ratio in the MDD-VHD algorithm increases from 0.00152

TABLE 5.5
The percentage of MDD-VHD reduction in packet loss ratio compared to Deswal in VoIP application

UE Group	HeNB	Reduction Percentage
10	20	95.0%
10	40	84.0%
10	60	88.0%
10	80	76.0%
20	20	92.0%
20	40	85.0%
20	60	77.0%
20	80	78.0%

to 0.0075 s when the HeNBs number increases from 20 to 80. In addition, the packet delay rate increases from 0.0215 to 0.0344 s with the same change in HeNBs number in the Deswal algorithm.

However, the proposed MDD-VHD algorithm reduces the packet loss ratio compared to the Deswal handover algorithm for each UE group and for each HeNBs density in the VoIP application as illustrated in Table 5.5.

As noted in Table 5.5, the of MDD-VHD algorithm reduces the packet loss ratio compared to the Deswal algorithm. Generally, the MDD-VHD algorithm in VoIP application reduces the packet loss ratio by 84.1% compared to Deswal handover algorithm in all scenarios.

However, the MDD-VHD algorithm shows the lowest packet loss in both applications compared to Deswal handover algorithm in each UEs group and in each HeNBs group. The reason as to why the MDD-VHD has a lower packet loss is attributed to the efficient use of UE trajectory through using the polynomial equation for the last four UE positions and ignoring all neighbor HeNBs that are not within the range of UE movement. In addition, the selection of the target HeNB on the basis of the closest distance and smallest θ further has a good effect on the reduction of redundant handover as each handover process has the probability of handover failure, and then is reflected on the packet loss ratio. Furthermore, it is also noted from the figure that there is an increase in the values of packet loss ratio in both algorithms when the UEs number was changed from 10 to 20 within each HeNBs density. This increment is predictable and is related to the increase in the density of UEs.

5.2.5 System Throughput

By increasing the HeNBs number from 20 to 80 in the three distribution scenarios (close, middle, and at the edge) using applications (CBR and VoIP), throughput decreases as shown in Figures 5.13 and 5.14. In both figures the effect of HeNBs number on system throughput is observed; all distribution scenarios show a decrement in the throughput.

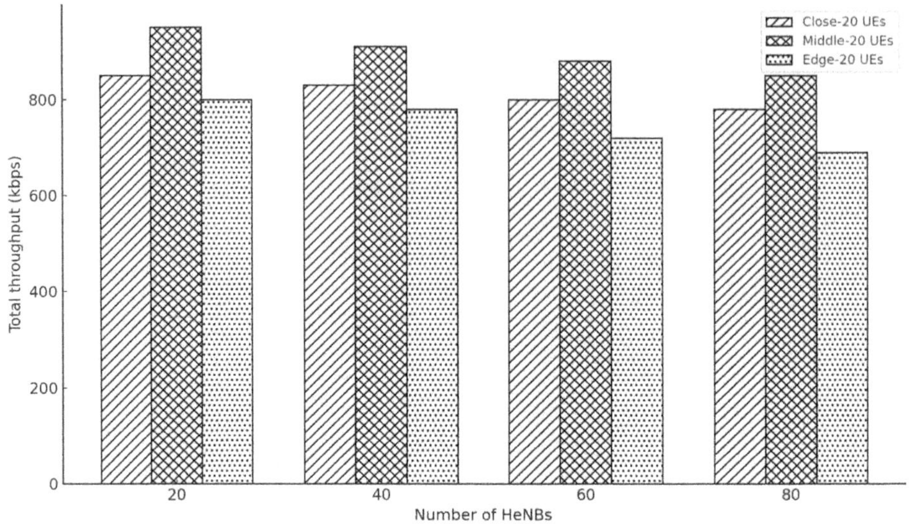

FIGURE 5.13 Comparison of system throughput of MDD-VHD algorithm in three scenarios using CBR application.

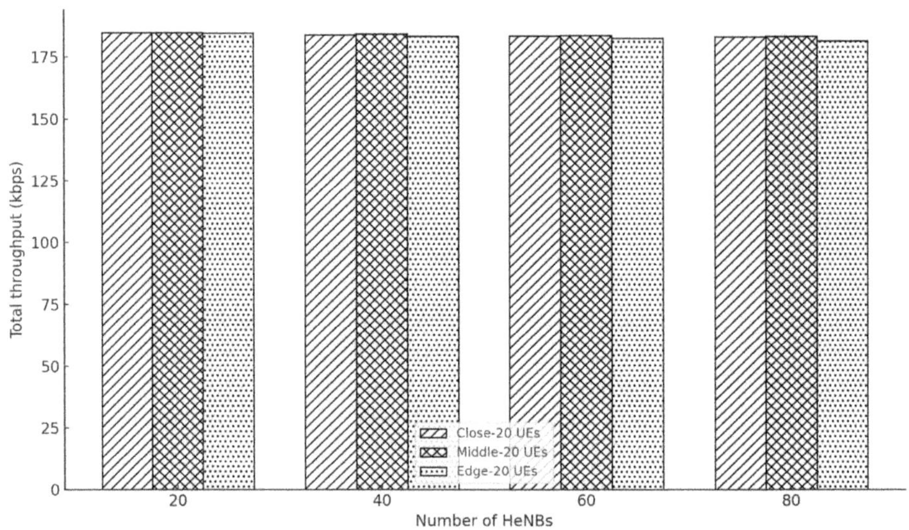

FIGURE 5.14 Comparison of system throughput of MDD-VHD algorithm in three scenarios using VoIP application.

Moreover, both figures show that the highest throughput is when the density of HeNBs is in the middle, while the lowest one is when the HeNB density is at the edge. This is due to the fact that the UE starts moving from the location of eNB tower and the density of HeNBs when concentrated close to the eNB tower affects the

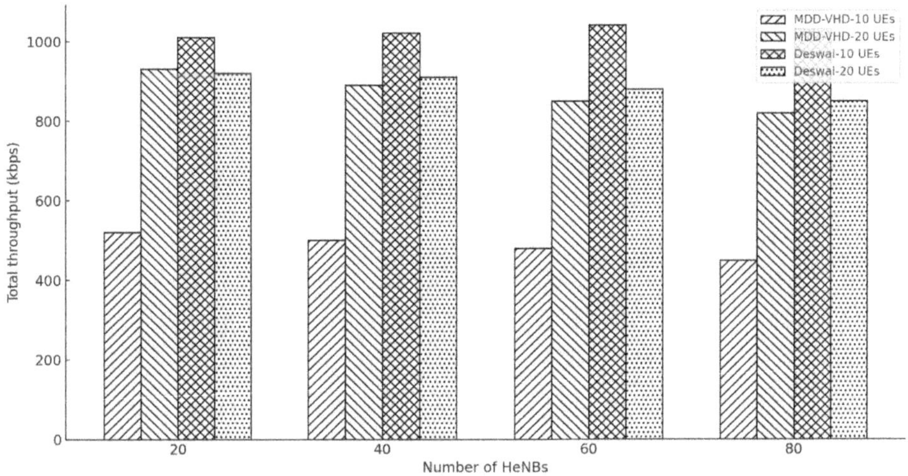

FIGURE 5.15 Comparison of system throughput using CBR application in two handover algorithms.

packet loss by increasing both the handovers number and interference. As a result, the increase in the packet loss ratio reduces the system throughput in both applications. Moreover, when HeNBs are distributed at the edge, the packet loss increases because of the weakness of RSS between the UE and its serving eNB while UE executes hand-out handover and the density of HeNBs is less than their density when distributed in the middle.

In addition, based on Figure 5.13, the variation in the system throughput is clear between the three distribution scenarios while the variation in the system throughput in Figure 5.14 is quite small. The reason is related to the different packet size and real-time priority between the two applications.

Regarding the comparison between the MDD-VHD and existing work in terms of throughput, two handover decision algorithms were evaluated using a different number of HeNBs distributions (20, 40, 60, and 80) for each UE group (10 and 20) using CBR application. In fact, as shown in Figure 5.15, the throughput of MDD-VHD algorithm is reduced from 503.86 to 480.50 kbps when HeNBs number increases from 20 to 80 and the number of UEs is 10, while in Deswal algorithm it is decreased from 488.11 to 395.52 kbps with the same change in the HeNBs number.

In addition, the same figure shows throughput results when the number of UEs is 20 with the same variation in HeNBs number. The throughput of MDD-VHD algorithm decreases from 1005.17 to 953.92 kbps when the HeNBs number increases from 20 to 80. Moreover, the throughput is decreased from 936.45 to 813.01 kbps with the same change in HeNBs number in Deswal algorithm.

On the contrary, the proposed MDD-VHD algorithm increases the system throughput in CBR application compared to the Deswal handover algorithm for each of the two UE groups and in each density of HeNBs as illustrated in Table 5.6.

From Table 5.6, it is observed that the MDD-VHD algorithm improves the system throughput compared to the Deswal algorithm. Overall, the MDD-VHD algorithm in

TABLE 5.6
The percentage of MDD-VHD improvement in system throughput compared to Deswal using CBR application

UE Group	HeNB	Reduction Percentage
10	20	3.2%
10	40	13.0%
10	60	22.0%
10	80	21.4%
20	20	7.2%
20	40	15.4%
20	60	15.2%
20	80	17.2%

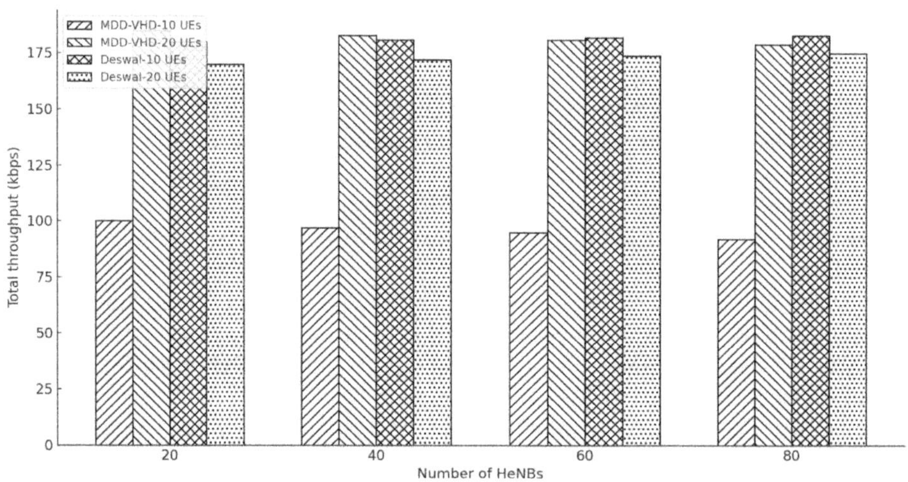

FIGURE 5.16 Comparison of system throughput using VoIP application in two handover algorithms.

CBR application increases the throughput by 14.3% compared to Deswal handover algorithm for all scenarios.

With regard to the VoIP application, the two handover decision algorithms were evaluated using a different number of HeNBs distributions (20, 40, 60, and 80) and a specific number of UES (10 and 20). Figure 5.16 illustrates the throughput of MDD-VHD and Deswal algorithms.

The throughput of MDD-VHD decreases from 92.68 to 91.43 kbps when the HeNBs number increases from 20 to 80 and the number of UEs is 10, while in Deswal algorithm it is decreased from 91.96 to 90.41 kbps with the same change in the HeNBs number. In addition, the same figure shows the throughput results when the number of

TABLE 5.7
The percentage of MDD-VHD improvement in system throughput compared to Deswal using VoIP application

UE Group	HeNB	Reduction Percentage
10	20	0.7%
10	40	1.0%
10	60	1.6%
10	80	1.1%
20	20	2.0%
20	40	2.1%
20	60	2.0%
20	80	2.8%

UEs is 20, with the same variation in HeNBs number. The throughput of MDD-VHD algorithm decreases from 185.13 to 184.15 kbps when the HeNBs number increases from 20 to 80. Furthermore, the throughput is decreased from 181.43 to 179.03 kbps with the same change in HeNBs number in Deswal algorithm.

On the other hand, the proposed MDD-VHD algorithm increases the system throughput using VoIP application compared to the Deswal algorithm for the two groups of UEs and in each density of HeNBs as illustrated in Table 5.7.

From Table 5.7, it is observed that the MDD-VHD algorithm improves the system throughput in comparison to the Deswal algorithm. In general, the MDD-VHD algorithm using VoIP application increases the throughput by 1.82% compared to Deswal handover algorithm in all scenarios.

However, the MDD-VHD algorithm shows the highest throughput in both applications compared to Deswal handover algorithm in each of the two UE groups. This is because the MDD-VHD has a lower packet loss and this is reflected on the total throughput. In addition, as noticed from the figures, the values of the system throughput decreased in both algorithms when the UEs number was changed from 10 to 20 within each HeNBs density. This increment can be predictable due to the increased density of UEs.

5.3 SUMMARY

The performance of the MDD-VHD algorithm and the threshold values for its parameters are investigated and validated in this chapter. Moreover, two applications (CBR and VoIP) are used to investigate the MDD-VHD performance in a different number of UEs and of HeNBs. In addition, three scenarios of HeNBs distribution are also considered in the performance of the proposed algorithm. The MDD-VHD algorithm shows the best performance with respect to the average handovers number, signaling measurements rate, packet delay rate, packet loss ratio, and system throughput compared to the benchmark handover decision algorithm as illustrated in all simulation scenarios. This improvement in the performance of the MDD-HVD algorithm is due to the prediction of future UE direction and the selection of the appropriate target HeNB from HeNB candidate list based on the closest distance and smallest angle. This list only includes the neighbor cells that fall within the range of future UE direction. Moreover, the MDD-VHD algorithm discusses all types of vertical handover: hand-in, hand-between, and hand-out. In hand-out handover, the MDD-VHD discusses two situations: returning back to home-serving eNB or connecting to the nearest eNB. In addition, the MDD-VHD algorithm performs much better than the handover algorithm of Deswal and Singhrova (2016). In addition, the discussion covers the effect of the MDD-VHD algorithm in the three cases of HeNBs distribution, when HeNBs spread near the eNB tower, when HeNBs spread in the middle of eNB coverage area, and when HeNBs spread at the edge of eNB coverage area. The results show an improvement in the system performance when the distribution of HeNBs is in the middle of the eNB coverage area.

6 Application of the Proposed Handover Algorithms in a Smart City Model

6.1 INTRODUCTION

Smart cities have emerged as a transformative concept in urban development, combining advanced technologies, communications networks, and infrastructure to create more efficient, sustainable, and livable environments as illustrated in Figure 6.1. These cities represent a dynamic ecosystem where millions of interconnected devices, sensors, and autonomous vehicles constantly exchange data to improve various services (Venkatesan et al. 2022). From real-time traffic management that reduces congestion to intelligent energy distribution systems that monitor consumption patterns, smart cities aim to improve the quality of life for their residents (Ahmad 2024). As smart cities expand, the demand for reliable, high-speed, and efficient communications networks has grown exponentially. This rapid growth has led to an increasingly complex landscape where multiple devices and applications coexist, requiring seamless connectivity to operate effectively. In this context, handoff management becomes a critical aspect, ensuring that UEs such as smartphones, sensors, and vehicles maintain uninterrupted connections as they move across different network cells. Handoff management in smart cities involves the seamless transition of connected devices between different network cells without loss of service or data. Given the densely packed environment of smart cities, where millions of devices constantly move and interact with the network, ensuring reliable connectivity is a significant challenge. Efficient handoff mechanisms are essential for maintaining QoS, particularly for latency-sensitive applications like autonomous driving, real-time surveillance, and emergency services (Abu-Ain et al. 2023). As devices shift between various network zones—each potentially operated by different communication technologies, such as 5G, Wi-Fi, and LPWAN—adaptive and intelligent handoff solutions are required to optimize bandwidth allocation, minimize packet loss, and reduce latency. Moreover, as urban populations grow and smart city infrastructure becomes more sophisticated, the demand for advanced, context-aware handoff protocols will only increase, further driving innovation in this field (Rafique et al. 2024).

In HetNets, which form the backbone of smart city communications infrastructure, frequent handovers between macrocells and small cells pose significant challenges in maintaining QoS. These challenges include managing increased handover frequency,

DOI: 10.1201/9781003648130-6

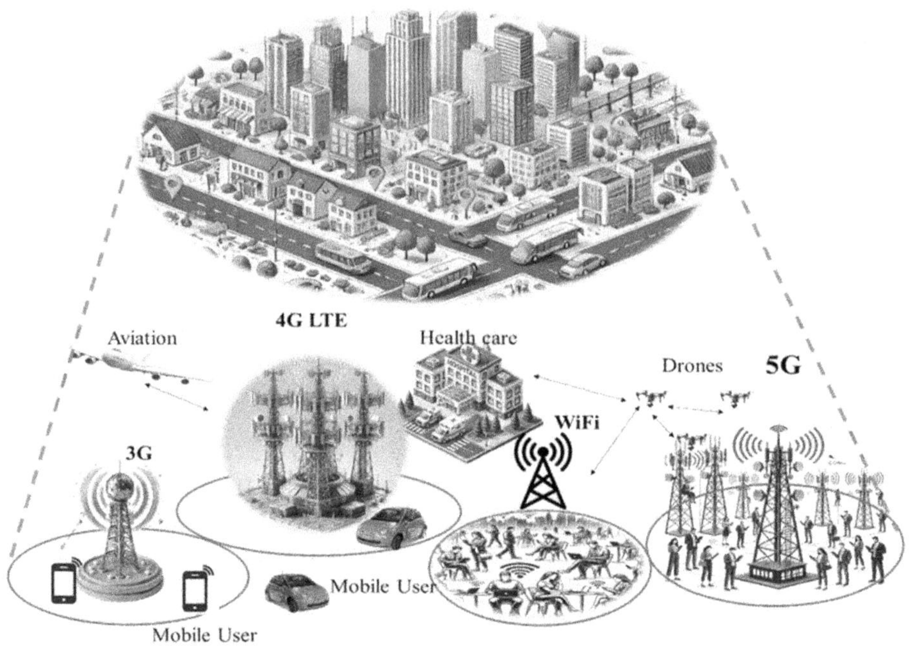

FIGURE 6.1 Advanced technologies, communications networks, and infrastructure.

avoiding unnecessary handovers, and reducing packet loss, especially in dense multi-cell urban environments where mobility patterns can be highly unpredictable. To address these challenges, advanced mobility management techniques such as predictive handover algorithms, machine learning-based decision-making, and context-aware handover strategies are being developed. These approaches aim to optimize the handover process by anticipating user movement, analyzing historical mobility patterns, and leveraging real-time environmental data. Additionally, the integration of edge computing and caching mechanisms helps reduce latency and improve data throughput during handovers, ensuring a more seamless user experience. By mitigating packet loss and unnecessary handovers, these innovations are crucial for maintaining high QoS in the complex, dynamic environments typical of smart cities. The movement (MDIH-HD and MDD-VHD) algorithms, as introduced in Chapters 3, 4, and 5, provide innovative solutions to address these challenges. By leveraging historical data, movement direction, and distance metrics, these algorithms improve handover decisions, thereby enhancing overall network efficiency and reducing the likelihood of outages. This chapter delves into the practical application of these algorithms within smart city models, demonstrating how they contribute to maintaining seamless connectivity for diverse applications in such dynamic and complex urban ecosystems.

6.2 THE ROLE OF LTE-A HETEROGENEOUS NETWORKS IN SMART CITIES

Heterogeneous LTE-A networks provide the foundation for supporting smart city applications due to their ability to provide high data rates, improved coverage, and efficient use of spectrum resources. The combination of eNBs and HeNBs allows for improved coverage both indoors and outdoors, ensuring that users and devices remain connected as they move through different environments. However, the high density of small cells in urban areas poses challenges, such as frequent handovers, increased interference, and signaling overhead (Fazio et al. 2012). For example, in a smart city scenario with heavy traffic and high mobility, user devices may pass through multiple small cells within minutes, resulting in multiple handovers. This situation can degrade the overall user experience, resulting in packet loss, delay, and reduced throughput. To mitigate these challenges, advanced techniques such as interference coordination, load balancing, and optimized handover algorithms are essential. Interference coordination strategies like enhanced Inter-Cell Interference Coordination help manage the interference between macro and small cells by allowing cells to dynamically adjust their transmission power and scheduling. Moreover, load balancing mechanisms ensure that the network resources are optimally distributed across cells, preventing congestion in high-traffic areas (Wang et al. 2020). In addition, the introduction of fast and predictive handover algorithms minimizes service interruptions and signaling overhead, enabling seamless connectivity for devices in high-mobility environments. These solutions collectively enhance the performance of heterogeneous LTE-A networks, making them better suited for the complex demands of smart city applications. Advantages of heterogeneous LTE-A networks in smart cities include the following:

- Improved Coverage: Large and small cells provide comprehensive coverage, which is essential for densely populated urban areas and high-rise buildings.
- High data rates: LTE-A supports applications that require high data rates, such as video surveillance and autonomous vehicle communications.
- Scalable and flexible architecture: The network can adapt to diverse user densities and mobility patterns, which is critical for smart city applications.

Challenges in handover management include the following:

- Frequent handover triggers: Due to dense deployment of small cells, frequent handovers are triggered, causing increased signal cost and potential packet loss.
- Interference management: Interference between large and small cells can affect the quality of handover decisions, leading to performance degradation.
- Handover failures: In scenarios with high UE mobility, handover failures can occur, leading to disconnections and reduced QoS.

The proposed MDIH-HD and MDD-VHD algorithms address these challenges by optimizing handover decisions, reducing unnecessary handovers, and improving overall network performance.

6.3 APPLICATION OF MDIH-HD AND MDD-VHD ALGORITHMS IN SMART CITIES

In smart cities, mobility scenarios vary widely due to the diverse range of environments and user behaviors. High-mobility scenarios often involve user units such as vehicles, public transportation, or drones moving at fast speeds across different areas, requiring frequent handovers between cells to maintain continuous connectivity. These scenarios are common in urban centers, where users are constantly on the move, and connectivity needs are heightened. On the other hand, low-mobility or stationary scenarios occur in places like residential or commercial areas where user units, such as IoT devices, sensors, or pedestrians, move slowly or remain static for extended periods of times. Both mobility scenarios present unique challenges; high mobility demands rapid handover and robust interference management, while low mobility requires stable, energy-efficient connections with minimal signaling. Managing these diverse scenarios effectively is key to maintaining a seamless and reliable network experience in a smart city.

The MDIH-HD and MDD-VHD algorithms are designed to handle different mobility scenarios in smart cities, ensuring that user units maintain seamless connectivity. These algorithms adapt to both high-mobility and low-traffic situations, which is critical to the diverse connectivity needs of smart city environments.

6.3.1 MDIH-HD Algorithm in High-Mobility Scenarios

In smart city environments, the high mobility of users, especially in densely populated areas, presents unique challenges for maintaining seamless connectivity. For instance, commuters on public transportation or fast-moving vehicles on highways may cross multiple small cells in a short period, requiring frequent handovers to maintain service quality. Additionally, pedestrians navigating urban environments may experience varying signal strengths as they move between buildings, open areas, and underground transit systems. Effectively managing handovers and maintaining a stable connection in such diverse and high-mobility scenarios is critical to ensuring reliable communication in smart cities (Olaverri-Monreal, 2016). The MDIH-HD algorithm excels in high mobility environments, such as highways, public transit routes, and urban arterial roads. By leveraging historical handover data and predicting the direction of user unit movement, the algorithm reduces unnecessary handovers and ensures accurate target cell selection. The MDIH-HD algorithm operates by analyzing patterns in user movement and handover history to predict future mobility paths. This predictive capability allows the algorithm to anticipate which cells a user is most likely to enter next cell, thereby reducing the number of handovers to cells. By focusing on selecting target cells that offer longer connection durations and stronger signal quality, the algorithm minimizes the signaling load on the network and reduces handover failures. In environments with high mobility, such as highways or urban roads with frequent turns and obstacles, this method significantly enhances network performance by providing more stable connections, reducing latency, and improving the overall user experience. Additionally, the algorithm adapts to real-time conditions, adjusting its predictions based on speed and direction to further refine its handover decisions. Detailed operation of MDIH-HD in smart cities:

- Using historical data: The MDIH-HD algorithm stores information about past deliveries, including the user unit's path, speed, and successful handovers. This data helps predict the most likely cell for the next handover, reducing unnecessary signals.
- Predictive direction analysis: By analyzing the direction of user device movement, the algorithm identifies cells in the user device's path and excludes cells that fall outside the direction of movement.

For example, consider an intelligent bus system moving along a predefined route with multiple eNBs and HeNBs along the route. The MDIH-HD algorithm can predict the next cell based on the historical route of the bus, greatly reducing handover failures and ensuring continuous Internet access for passengers. Advantages in high-mobility scenarios include the following:

- Reduced handover failures: The algorithm reduces the probability of selecting the wrong target cell by considering the orientation of the user device.
- Reduced signaling costs: By filtering out irrelevant cells, the signals required for handover decisions are reduced, enhancing network efficiency.

6.3.2 MDD-VHD ALGORITHM IN LOW TRAFFIC SCENARIOS

In heterogeneous network environments, particularly those involving a mix of macro and small cells, efficient handover mechanisms are crucial to maintaining seamless connectivity as users move across different coverage areas. These handovers need to be optimized to minimize latency and packet loss, especially in scenarios where user devices frequently move between varying network layers (Iftikhar et al. 2023). Traditional handover algorithms primarily focus on signal strength and mobility patterns but may not always deliver optimal performance in low traffic environments where different dynamics, such as user orientation and localized traffic conditions, come into play. This necessitates the development of more sophisticated handover algorithms like the MDD-VHD, which is designed to address the unique challenges of low traffic areas. The MDD-VHD algorithm is specifically designed for low traffic scenarios, such as pedestrian areas, shopping malls, and residential areas. This algorithm considers the UE orientation, distance to the target cell, and signal strength to make accurate vertical handover decisions. By considering the orientation of the UE, the algorithm can predict the movement direction, enabling proactive handover decisions. Additionally, it assesses the distance between the UE and the target cell, ensuring that handovers occur only when the UE moves closer to a stronger signal source. The algorithm also evaluates the signal strength from both the serving and target cells to prevent unnecessary handovers, minimizing signal interruptions and improving the overall user experience in areas with slower-moving traffic or more static users, such as residential neighborhoods or shopping malls. This results in more efficient handover processes, reducing signaling overhead and ensuring stable connectivity in low-mobility environments. Detailed operation of MDD-VHD in smart cities includes the following:

- Distance measurement: The algorithm measures the distance between the UE and the candidate cells, prioritizing the cells that are closest and have the strongest signal.
- Direction analysis: Similar to MDIH-HD, the MDD-VHD algorithm uses the direction of movement of the UE to determine the most suitable small cell for handover.

For example, in a smart energy management system with connected smart meters in residential buildings, the MDD-VHD algorithm ensures stable connectivity by selecting the most suitable small cell, even when users move around their homes. Advantages in low traffic scenarios include the following:

- Reduced packet loss: By selecting the closest cell with the strongest signal, the algorithm reduces the chances of packet loss during handover.
- Efficient use of network resources: The algorithm ensures that user units remain connected to the most suitable cell, reducing unnecessary handovers.

6.3.3 Optimizing Vertical Handover for Smart IoT Applications

In smart IoT applications, vertical handover is essential for maintaining seamless connectivity as devices transition between different types of network cells, such as macro cells and small cells. Low-mobility IoT devices, including smart meters, environmental sensors, and surveillance cameras, often operate in fixed or slow-moving locations (Khattab et al. 2023). Despite their limited mobility, these devices still require uninterrupted connections to ensure continuous data flow and support smart city services. The MDD-VHD algorithm is particularly valuable in such scenarios because it allows for efficient handover decisions based on factors like signal strength and the proximity of the device to the nearest cell. This ensures that IoT devices remain connected to the optimal network, reducing interruptions and improving the reliability of critical applications in smart cities. The MDD-VHD algorithm plays a crucial role in managing handovers between different network cells, especially in scenarios involving low-mobility IoT devices such as smart meters, environmental sensors, and surveillance cameras. These devices require stable and consistent connectivity to ensure reliable data transmission, which is essential for maintaining smart city services.

- Improving data transmission efficiency: The MDD-VHD algorithm improves data transmission efficiency by reducing packet loss and ensuring that IoT devices remain connected to the ideal cell, even in low-mobility situations. This optimization is particularly vital for applications that rely on real-time data, such as air quality monitoring and smart street lighting.

- Ensuring the reliability of smart city services: By intelligently managing the handover process, the algorithm contributes to the reliability and stability of smart city services, preventing outages that could compromise the functionality of critical systems.

6.3.4 Addressing Network Congestion and Interference

In smart IoT environments, network congestion and interference are critical issues, particularly in densely populated urban areas where a vast number of connected devices generate constant data traffic. Smart cities face challenges of network congestion and interference, especially in dense urban environments with high concentration of user devices. Congestion occurs when multiple IoT devices, such as sensors, cameras, and smart appliances, simultaneously attempt to transmit data over the same network, overwhelming the available bandwidth and causing delays, packet loss, or even network outages (Pons et al. 2023). Interference, on the other hand, arises when signals from closely located devices or networks operating on overlapping frequency bands clash, degrading signal quality and leading to reduced communication reliability. In a smart city, where IoT devices are widely deployed for applications like traffic management, environmental monitoring, and public safety, these challenges can significantly hamper the system's performance, making it difficult to ensure real-time data transmission and seamless device-to-device communication. Efficient congestion control mechanisms and interference management strategies are crucial to maintaining the reliability and scalability of IoT networks in such scenarios. The proposed network effectively addresses these challenges by leveraging both macrocells (eNB) and femtocells (HeNB), ensuring seamless connectivity even in congested areas.

- Dealing with interference in dense urban environments: The combination of MDIH-HD and MDD-VHD algorithms helps manage interference by ensuring that devices connect to the most suitable cells based on their movement patterns. Both techniques rely on analyzing both the direction and distance of user devices, help manage interference by ensuring that devices connect to the most suitable cells based on their mobility patterns. By considering the orientation and proximity of user equipment, the system can effectively filter out less relevant cells, which is particularly useful during large public events or in high-traffic zones where network congestion is common.
- Improving signal quality: By focusing on direction and distance metrics, along with the use of macrocells and femtocells, the proposed algorithm improves signal quality by reducing interference. This results in fewer dropped connections and improved quality of service for end users, even in densely populated smart city environments.

6.4 CASE STUDIES: SMART CITY APPLICATIONS

The application of MDIH-HD and MDD-VHD algorithms in smart city environments demonstrates their ability to enhance handover efficiency and maintain seamless connectivity. The following case studies illustrate the practical impact of these algorithms across different smart city applications, emphasizing how they address the unique challenges posed by high mobility, dense network cell deployment, and diverse communication needs.

6.4.1 Intelligent Transportation Systems (ITS)

- Scenario: In ITS, particularly in urban areas with high-speed public transportation like buses, trams, and self-driving vehicles, frequent handovers are a significant challenge due to the continuous movement of these vehicles across multiple network cells. As these vehicles travel at high speeds, the user devices and onboard communication systems must constantly switch between different base stations or cells, which can lead to service interruptions, latency, and increased signaling overhead if not managed efficiently (Rani & Sharma 2024). In such scenarios, maintaining seamless connectivity for real-time applications like GPS navigation, traffic updates, or vehicle-to-vehicle communication is critical. Without optimized handover algorithms, users may experience dropped connections, delayed data transmission, and poor overall service quality. Figure 6.2 shows the emerging ITS infrastructure.
- MDIH-HD algorithm application: The MDIH-HD algorithm effectively predicts the movement patterns of highly mobile UE, such as those in buses and trams. By leveraging historical data and movement direction, the algorithm ensures that handoffs occur smoothly as vehicles move between macrocells and microcells. This reduces unnecessary handoffs and ensures

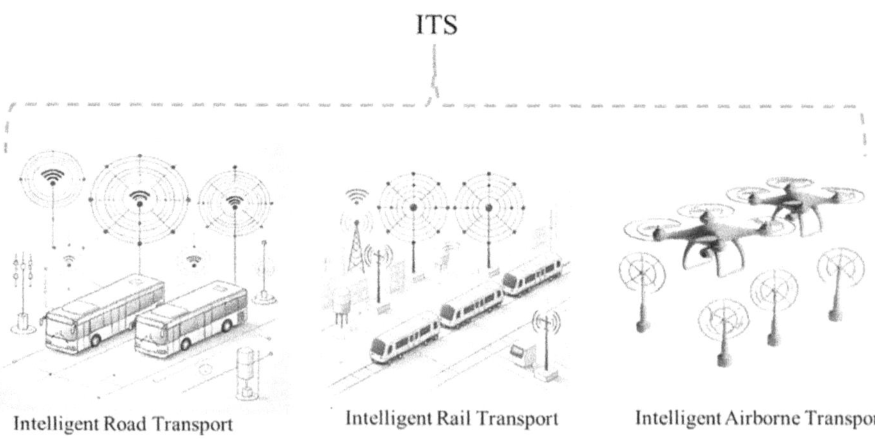

FIGURE 6.2 Emerging ITS infrastructure.

stable connectivity for services such as real-time navigation, passenger Wi-Fi, and safety monitoring systems.

- Result: Reduced packet loss and signal overhead, improved handoff success rates, and improved passenger experience, especially during peak times.

6.4.2 Smart Healthcare Systems

- Scenario: In the context of smart healthcare services, stable and reliable connectivity is critical for ensuring the continuous monitoring and timely response to patient needs (Ahmad et al. 2023). For instance, remote patient monitoring systems rely on a network of sensors and wearable devices to collect and transmit vital signs such as heart rate, blood pressure, and oxygen levels to healthcare providers in real time (Peralta-Ochoa et al. 2023). In emergency situations, such as heart attacks or accidents, seamless connectivity is essential for the immediate transmission of medical data and live video feeds to emergency response teams, enabling them to make quick, informed decisions. In smart cities, where these services are deployed across both residential and commercial areas, handover efficiency becomes crucial, especially as patients or healthcare personnel move between different network cells (Figure 6.3).
- Application of MDD-VHD algorithm: In low-traffic scenarios, such as patients using home health monitoring devices, MDD-VHD algorithm ensures stable connections by selecting the most suitable small cell based on proximity and signal strength. This reduces packet loss and improves service quality, which is critical for timely data transmission and emergency alerts.
- Result: Enhance the reliability of healthcare services, reduce handover failures, and optimize resource utilization for smart healthcare applications.

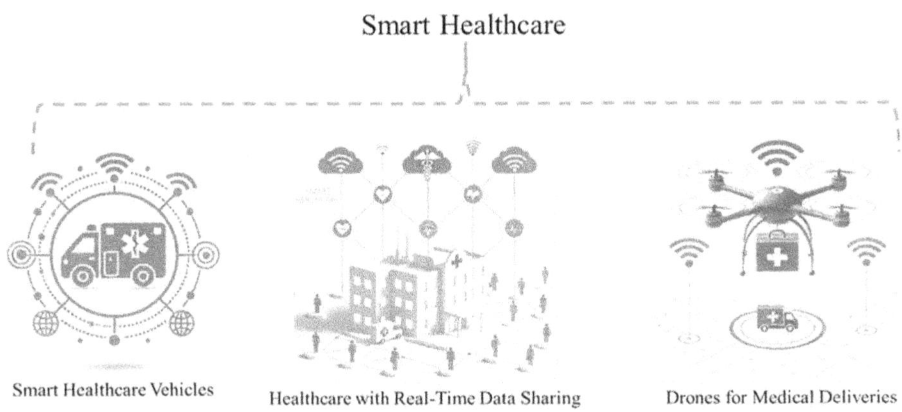

FIGURE 6.3 Emerging Smart Healthcare infrastructure.

6.4.3 Autonomous Vehicle Communication

- Scenario: In emerging concept of smart cities, autonomous vehicles must navigate complex urban environments, constantly interacting with infrastructure, traffic signals, and other vehicles. A key challenge in this dynamic scenario is seamless handover management, ensuring that vehicles maintain uninterrupted connectivity as they move through different communication cells or zones. Autonomous vehicles rely heavily on real-time data exchange with surrounding infrastructure and other vehicles for making safe and efficient driving decisions (Hakak et al. 2023). For instance, when a vehicle moves at high speeds through a city, it needs to maintain consistent communication with nearby traffic lights, road sensors, and neighboring vehicles to adapt to changing conditions in real time, such as traffic congestion or obstacles.
- MDIH-HD algorithm application: For autonomous vehicles moving at high speeds, the MDIH-HD algorithm effectively predicts direction and movement patterns, ensuring that deliveries are made to the most appropriate cells without delay. This is critical to maintaining real-time data exchange, ensuring safety, and enhancing autonomous decision-making.
- Result: Improved safety and reliability in autonomous vehicle operations, reduced handover latency, and consistent vehicle-to-infrastructure and vehicle-to-vehicle communications.

6.4.4 Smart Energy Management

- Scenario: In smart cities, energy management relies on the seamless transmission of consumption data from smart meters and IoT devices to energy providers. These devices, which are often low-mobility and stationary, need to send real-time data on electricity usage to optimize energy distribution, monitor consumption patterns, and ensure accurate billing (Mishra & Singh 2023). The challenge in such scenarios lies in maintaining stable, efficient communication, especially as devices may face interference or connection instability due to urban density or environmental factors.
- Application of MDD-VHD algorithm: In this context, MDD-VHD algorithm ensures that smart meters and other low-mobility IoT devices maintain stable communication with the nearest small cell. The algorithm's use of distance and direction metrics ensures accurate handover decisions, reducing packet loss and ensuring efficient data transmission.
- Result: Reliable data collection for energy providers, reduced packet loss, improved billing accuracy, and energy consumption monitoring.

6.4.5 Public Safety and Emergency Services

- Scenario: In the context of public safety and emergency services, maintaining real-time communication is critical for ensuring public safety and enabling rapid emergency response. A key challenge in this scenario is the seamless transmission of data from various sources, including surveillance cameras, IoT sensors, and mobile units like police vehicles. These data streams are essential for situational awareness, but they come with significant challenges. One of the primary challenges is managing high mobility environments where mobile units frequently change locations (Damaševičius et al. 2023). Police vehicles, for instance, often travel through areas with varying network coverage. In such cases, ensuring uninterrupted communication without delays, dropped signals, or data loss is difficult, especially when video streams and sensor data must remain continuous and reliable. Additionally, fixed surveillance devices, such as cameras installed in city streets or public buildings, generate vast amounts of data that need to be transmitted in real time (Chukhno et al. 2023). These devices are typically fixed in high-traffic areas where network congestion may occur. Managing the high data volumes and maintaining stable connections in such environments poses another significant challenge, as any disruption could hinder public safety efforts.

- Application of both MDIH-HD and MDD-VHD algorithms: Mobile surveillance units, such as police vehicles, rely on the MDIH-HD algorithm to efficiently manage handover during high mobility, while fixed devices such as surveillance cameras leverage the MDD-VHD algorithm to maintain stable communications. Both algorithms ensure seamless data transfer, allowing law enforcement agencies and emergency responders to access real-time data without interruption.
- The result: Improved situational awareness, faster response times during emergencies, and enhanced public safety through continuous connectivity.

6.5 SUMMARY

This chapter provided a comprehensive examination of how the proposed handover algorithms, MDIH-HD and MDD-VHD, can be effectively applied within a smart city model to enhance network performance and maintain seamless connectivity. It introduced smart cities as transformative urban ecosystems integrating advanced technologies, communication networks, and infrastructure to create more efficient, sustainable, and livable environments characterized by millions of interconnected devices, sensors, and autonomous vehicles exchanging data to improve services like real-time traffic management and intelligent energy distribution. The chapter highlighted the challenges in handover management within smart cities, such as frequent handover triggers due to dense small cell deployment, interference management between macro and small cells, and handover failures caused by high UE mobility, which can lead to increased signaling costs, packet loss, and reduced QoS. It emphasized the role of LTE-A heterogeneous networks HetNets in supporting smart city applications by providing improved coverage, high data rates, and scalable architecture, while acknowledging the need for advanced techniques like interference coordination, load balancing, and optimized handover algorithms to address the challenges posed by dense deployment of small cells.

The application of the MDIH-HD and MDD-VHD algorithms was discussed in the context of managing different mobility scenarios in smart cities to ensure seamless connectivity for UEs. The MDIH-HD algorithm leverages historical data and predicts movement direction in high-mobility scenarios like highways and public transit routes, reducing unnecessary handovers and signaling load, while the MDD-VHD algorithm considers UE orientation, distance to target cells, and signal strength in low-mobility scenarios such as pedestrian areas and residential zones, ensuring accurate vertical handover decisions and stable, energy-efficient connections. The chapter also highlighted how these algorithms optimize vertical handovers for smart IoT applications by enhancing data transmission efficiency and reliability of services, and addressed network congestion and interference by managing interference through intelligent cell selection based on direction and distance metrics, improving signal quality, and reducing dropped connections. Practical case studies illustrated the impact of these algorithms across various smart city applications: in ITS, the MDIH-HD algorithm reduces packet loss and improves handover success rates for high-speed public transportation; in smart healthcare systems, the MDD-VHD algorithm enhances reliability and reduces handover failures for remote patient monitoring; in autonomous vehicle communication, the MDIH-HD algorithm improves safety and reduces handover latency; in smart energy management, the MDD-VHD algorithm ensures reliable data collection and reduces packet loss for stationary IoT devices; and in public safety and emergency services, both algorithms improve situational awareness and response times by ensuring continuous connectivity. The chapter concluded by underscoring the critical role of these advanced handover algorithms in maintaining seamless connectivity and enhancing network performance in smart cities, effectively addressing challenges posed by high mobility, dense network deployments, and diverse communication needs, thereby contributing significantly to the reliability and effectiveness of smart city services.

7 Conclusions and Key Takeaways

7.1 FINDINGS

The LTE-A system was developed to fulfill the demand for increased bandwidth and higher communication velocity. Thus, this system is equipped with new techniques and algorithms that improve system performance and user experience. However, various concerns about the suitability and adaptability of the existing handover decision algorithms for LTE-A systems have arisen due to the new technologies and methods introduced. Nevertheless, such new techniques encourage the development of new methods and algorithms to enhance the overall system performance in general.

Therefore, this book focuses on the LTE-A system handover decision algorithms that will provide enhancements to the handover experience and performance in LTE-A HetNet. Moreover, the scope of this study was expanded to include an investigation into the challenges related to the handover procedure. Hence, this study is motivated by the fundamental problem of increasing the number of wireless towers in the same geographical area to support the growing demands of wireless networks. Thus, frequent handovers among cells reduce network performance. Based on the solutions proposed recently in several studies, an investigation was conducted on the possibility of using a GPS service to enhance the new efficient handover decision procedure. Using the GPS system, future UE movements can be predicted. This is achieved through the detection of UE trajectory, cell location, and the distance between them. This characteristic helps UEs to maintain a steady connection through the LTE-A system in both types of network cells (eNB and HeNB). The MDIH-HD algorithm is used when the UE mobility velocity is more than 30 km/h, whereas the eNB network cell is only used for the handover procedure. Moreover, the target eNB is selected based on the prediction of future UE location when the RSS of its serving eNB is lower than the predefined threshold value. Otherwise, the target eNB is selected directly from the UE table if the UE uses a routine route. Based on reducing the searching time of the target eNB, especially at high UE mobility velocities, the handover loss and delay is improved and the system performance is increased.

Furthermore, a new MDD-VHD algorithm was also proposed for the vertical handover between two-tier LTE-A networks (eNB and HeNB). The MDD-VHD is used when the UE mobility velocity is less than 30 km/h, as is the case for pedestrians or cyclists. This is to reduce the load of eNB and improve the QoS for users. The

DOI: 10.1201/9781003648130-7

MDD-VHD algorithm uses the UE trajectory, which depends on a polynomial function, to predict user future movement with appropriate target HeNB selection. This is done using the distance between the current UE position and HeNB location. In addition, the target HeNB should be within the range of UE directions to improve the LTE-A system performance throughput, which is done by reducing redundant handovers.

The two proposed algorithms were implemented and simulated using an LTE-Sim simulator. In addition, both of the proposed algorithms and their experimental results were divided and presented separately in two chapters in this book. In Chapter 6, the MDIH-HD algorithm, its experimental results, and performance evaluation are presented and discussed. Moreover, the MDIH-HD algorithm is compared with the 3GPP standard and the handover algorithm of Wang et al. (2014a). In Chapter 5, the MDD-VHD algorithm, its experimental results, and performance evaluation are discussed and presented. Furthermore, the MDD-VHD algorithm is compared with the algorithm of Deswal and Singhrova (2016) for which the MDD-VHD algorithm is applied in three distribution scenarios (close, middle, and edge).

The results obtained from each of the proposed algorithms indicate a higher percentage of success in data delivery compared to the benchmarks used in this work. The performance metrics selected in each algorithm are the average handover number, average signaling measurement number, packet delay rate, packet loss ratio, and system throughput. Furthermore, the packet delay and loss of both algorithms reveal valuable improvement in the different scenarios.

7.2 CONTRIBUTIONS

The main contribution of this book is the proposal of efficient handover algorithms for application in horizontal and vertical networks in the LTE-A system. The newly proposed handover techniques show promise in reducing packet delay, packet loss, unnecessary handover numbers, and measurement of corresponding messages. At the same time, the wireless network throughput is also improved. Briefly, the contributions of this work are listed as follows:

1. A prioritized multi-criterion parameter selection algorithm that takes into consideration mobility and system efficiency is developed. Additional parameters, such as distance, UE trajectory, cell location, and UE velocity, are used to complement the RSS and TTT of existing handover decision parameters. This new combination of multi-parameters successfully mitigates the weakness of the conventional handover procedure in both horizontal and vertical networks.
2. A fast target cell determination mechanism is introduced through the adoption of the proposed multi-criterion parameter selection algorithm to cater to cell-edge users. An MDIH-HD algorithm is introduced to reduce the searching time for selecting target cells when the UE velocity is high. This

reduction is based on the use of a UE mobility history table in the routine route mobility or according to UE movement direction when the UE moves in a new pathway.

3. Efficient switching between horizontal and vertical handover due to low mobility is achieved. A MDD-VHD algorithm is presented to select the appropriate target cell and reduce the searching time for the target cell based on the UE profile and HeNB cell positions when the UE velocity is low.

7.3 LIMITATIONS

The simulation results show that the MDIH-HD and MDD-VHD algorithms perform well in enhancing the handover decision in LTE-A networks. In the LTE-A architecture of this work, the proposed MDIH-HD algorithm on eNB cells was discussed to improve the handover decision procedure in both mobility models, that is, the random direction and routine route at high UE velocity. If the UE moves in the routine route and its trajectory is saved in the table, the target eNB will be selected from the table. This is because it is supposed that all eNB cells should accept the new UE. In fact, when the capacity of the target eNB is full, handover failure will occur. The reason being that the target eNB will reject accepting the new UE. Moreover, in the MDD-VHD algorithm, the capacity parameter of HeNB cell was not considered in selecting the target cell.

In addition, there are three types of HeNB (open, close, and hybrid). The simulation results examined the proposed MDD-VHD algorithm for HeNB cells of the open access type in order to enhance the target HeNB selection in the vertical handover decision. Therefore, using the other two types of HeNB cells would greatly influence the target HeNB selection and thus affect the handover decision procedure.

Furthermore, the UE mobility behavior is a crucial parameter in the handover decision procedure that will greatly affect the stability of the selected path. In fact, based on previous studies, their implementation depends on the fixed velocities of UE mobility. Therefore, for better improvement of handover performance, the acceleration in UE velocities cannot be neglected.

7.4 RECOMMENDATIONS

Utilizing LTE-A system in the handover decision parameters and improving the handover decision efficiency has proven effective. Thus, such parameters are, evidently, essential to improving different network operations. However, the selection of suitable parameters to improve the handover decision still encompasses a wide area of research. Therefore, the recommendation for further research in this field can be summarized as follows:

1. Investigating different parameters of user performance in light of handover and load balance in the wireless system over horizontal and vertical networks.

2. Investigating different parameters of user performance on both HeNB types, the close and hybrid.
3. With regard to implementation, UE velocity should be taken into account in the handover decision when switching between MDIH-HD and MDD-VHD algorithms as the main behavior. Thus, by monitoring the three main behaviors of UE, which are the UE mobility, acceleration, and deceleration as the frequent line changes, the suitability of each of the proposed algorithms (MDIH-HD and MDD-VHD) for the UE behavior can be ensured.

References

3GPP. 2007. Universal Mobile Telecommunications System (UMTS); Radio Resource Management Strategies (Release 7). *3rd Generation Partnership Project TR 25.922* 7.1.0: 0–97.

3GPP. 2010a. Evolved Universal Terrestrial Radio Access (E-UTRA); Base Station (BS) Radio Transmission and Reception (Release 9). *3rd Generation Partnership Project TS 136 104* 9.4.0: 0–99.

3GPP. 2010b. Technical Specification Group Radio Access Network; Evolved Universal Terrestrial Radio Access (E-UTRA); Further Advancements for E- UTRA Physical Layer Aspects (Release 9). *3rd Generation Partnership Project TR 36.814* 9(3): 1–104.

3GPP. 2011a. Digital cellular telecommunications system (Phase 2+); Universal Mobile Telecommunications System (UMTS); LTE; Vocabulary for 3GPP Specifications (Release 10). *3rd Generation Partnership Project TR 21.905* 10.3.0: 0–56.

3GPP. 2011b. Evolved Universal Terrestrial Radio Access (E-UTRA); Physical Channels and Modulation (Release 10). *3rd Generation Partnership Project TS 136 211* 10.1.0: 0–105.

3GPP. 2012a. LTE; Evolved Universal Terrestrial Radio Access (E-UTRA); Radio Resource Control (RRC); Protocol Specification (Release 10). *3rd Generation Partnership Project TS 136 331* 10.7.1: 0–306.

3GPP. 2012b. Technical Specification Group Radio Access Network; Evolved Universal Terrestrial Radio Access (E-UTRA) and Evolved Universal Terrestrial Radio Access Network (E-UTRAN); Overall Description; Stage 2 (Release 11). *3rd Generation Partnership Project TS 36 300* 11.4.0: 0–208.

3GPP. 2012c. Technical Specification Group Radio Access Network; Evolved Universal Terrestrial Radio Access (E-UTRA); Mobility Enhancements in heterogeneous networks (Release 11). *3rd Generation Partnership Project TR 36.839* 11.0.0: 1–53.

3GPP. 2012d. Telecommunication Management; Automatic Neighbour Relation (ANR) Management; Concepts and Requirements (Release 11). *3rd Generation Partnership Project TS 32.511* 11.2.0: 0–15.

3GPP. 2013. Evolved Universal Terrestrial Radio Access (E-UTRA) and Evolved Universal Terrestrial Radio Access Network (E-UTRAN); Overall Description; Stage 2 (Release 11). *3rd Generation Partnership Project TS 136 300* 11.5.0: 1–223.

3GPP. 2014a. LTE; Feasibility study for Further Advancements for E-UTRA (LTE- Advanced) (Release 12). *3rd Generation Partnership Project TR 136 912* 12.0.0: 1–61.

3GPP. 2014b. Overview of 3GPP Release 10.www.3gpp.org/specifications/ releases/70-release-10 [11-05-2017].

3GPP. 2014c. Overview of 3GPP Release 11. www.3gpp.org/specifications/ releases/69-release-11 [11-05-2017].

3GPP. 2014d. Overview of 3GPP Release 12. www.3gpp.org/specifications/ releases/68-release-12 [11-05-2017].

3GPP. 2014e. Technical Specification Group Services and System Aspects, Architecture Enhancements for non-3GPP Accesses (Release 12). *3rd Generation Partnership Project TS 23.402* 12.5.0: 1–80.

3GPP. 2014f. Technical Specification Group Services and System Aspects, General Packet Radio Service (GPRS) Enhancements for Evolved Universal Terrestrial Radio Access Network (E-UTRAN) Access (Release 12). *3rd Generation Partnership Project TS 23.401* 12.5.0: 1–89.

3GPP. 2014g. Universal Mobile Telecommunications System (UMTS); UTRAN Functions, Examples on Signalling Procedures (Release 12). *3rd Generation Partnership Project TR 25.931* 12.0.0: 0–110.

3GPP. 2015a. Digital cellular telecommunications system (Phase 2+); Universal Mobile Telecommunications System (UMTS); LTE; Policy and Charging Control Architecture (Release 12). *3GPP TS 23.203* 12.7.0: 0–125.

3GPP. 2015b. LTE; Evolved Universal T l Terrestrial Radio Access (E-UTRA); Physical c l channels and modulation (Release 12). *3rd Generation Partnership Project TS 1 136 211* 12.5.0: 0–292.

3GPP. 2015c. LTE; Evolved Universal Terrestrial Radio Access (E-UTRA); LTE physical layer; General description (Release 12). *3rd Generation Partnership Project TS 36 201* 12.2.0: 0–292.

3GPP. 2016a. Evolved Universal Terrestrial Radio Access (E-UTRA); Radio Frequency (RF) system scenarios (Release 13). *3rd Generation Partnership Project TR 36 942* 13.0.0: 0–84.

3GPP. 2016b. LTE; Evolved Universal Terrestrial Radio Access (E-UTRA) and Evolved Universal Terrestrial Radio Access Network (E-UTRAN); Overall description; Stage 2 (Release 13). *3rd Generation Partnership Project TS 136 300* 13.2.0: 0–299.

3GPP. 2016c. LTE; Evolved Universal Terrestrial Radio Access (E-UTRA); Physical layer; Measurements (Release 13). *3rd Generation Partnership Project TS 136 214* 13.0.0: 0–15.

Abu-Ain, T., Ahmad, R., Wazirali, R. & Abu-Ain, W. 2023. A new SDN-handover framework for QoS in heterogeneous wireless networks. *Arabian Journal for Science and Engineering*, 48(8), 10857–10873.

Ahmad, R. 2024. Smart remote sensing network for disaster management: An overview. *Telecommunication Systems*, 87, 213–237. https://doi.org/10.1007/s11235-024-01148-z

Ahmad, R., Hämäläinen, M., Wazirali, R. & Abu-Ain, T. 2023. Digital-care in next generation networks: Requirements and future directions. *Computer Networks*, 224, 109599.

Ahmad, R., Sundararajan, E.A. & Khalifeh, A. 2020. A survey on femtocell handover management in dense heterogeneous 5G networks. *Telecommunication Systems*, 75(4), 481–507.

Akyildiz, I.F., Gutierrez-Estevez, D.M. & Reyes, E.C. 2010. The evolution to 4G cellular systems: LTE-Advanced. *Physical Communication,* 3(4), 217–244.

Almulla, M., Wang, Y., Boukerche, A. & Zhang, Z. 2014. Design of a fast location- based handoff scheme for IEEE 802.11 vehicular networks. *IEEE Transactions on Vehicular Technology,* 63(8), 3853–3866.

Anas, M., Calabrese, F.D., Mogensen, P.E., Rosa, C. & Pedersen, K.I. 2007. Performance evaluation of received signal strength based hard handover for UTRAN LTE. *IEEE Vehicular Technology Conference*, Dublin, Ireland, pp. 1046–1050.

Balan, I., Sas, B., Jansen, T., Moerman, I., Spaey, K. & Demeester, P. 2011. An enhanced weighted performance-based handover parameter optimization algorithm for LTE networks. *EURASIP Journal on Wireless Communications and Networking,* 2011(1), 1–11.

Bangerter, B., Talwar, S., Arefi, R. & Stewart, K. 2014. Networks and devices for the 5G era. *IEEE Communications Magazine,* 52(2), 90–96. https://doi.org/10.1109/MCOM.2014.6736748

Belloni, F. 2004. Fading models. S-88 Signal Processing Laboratory, HUT (November), pp. 1–25.

Ben-Mubarak, M.A., Ali, B.M., Noordin, N.K., Ismail, A. & Ng, C.K. 2013. Mobile station movement direction prediction (MMDP) based handover scanning for mobile WiMAX system. *Wireless Personal Communications,* 73(3), 839–865.

Bilen, T., Duong, T.Q. & Canberk, B. 2016. Optimal eNodeB estimation for 5G intra- macrocell handover management. *Proceedings of the 12th ACM Symposium on QoS and Security for Wireless and Mobile Networks – Q2SWinet '16*, Malta, pp. 87–93.

Bononi, L., Bracuto, M., D'Angelo, G. & Donatiello, L. 2004. Performance analysis of a parallel and distributed simulation framework for large scale wireless systems. *Proceedings of the 7th ACM international symposium on Modeling, analysis and simulation of wireless and mobile systems – MSWiM '04*, p. 52.

Capozzi, F., Piro, G. & Grieco, L. 2012. A system-level simulation framework for LTE femtocells. *Proceedings of the 5th International ICST Conference on Simulation Tools and Techniques*, Desenzano, Italy, pp. 6–8.

Chandavarkar, B.R. & Guddeti, R.M.R. 2016. Simplified and improved multiple attributes alternate ranking method for vertical handover decision in heterogeneous wireless networks. *Computer Communications,* 83, 81–97.

Chang, C.W., Lin, Y.H., Jan, R.H. & Chen, C. 2012. Efficient measurement procedure for handover in LTE femtocell networks. *2012 International Conference on Selected Topics in Mobile and Wireless Networking, ICOST 2012,* pp. 119–123.

Chang, F.M., Wang, H.L., Hu, S.Y. & Kao, S.J. 2013. An efficient handover mechanism by adopting direction prediction and adaptive time-to-trigger in LTE networks. *Computational Science and Its Applications – ICCSA 2013, Pt V*, 7975, 270–280.

Chang, J., Li, Y., Feng, S., Wang, H., Sun, C. & Zhang, P. 2009. A fractional soft handover scheme for 3GPP LTE-Advanced system. *2009 IEEE International Conference on Communications Workshops*, pp. 1–5.

Chatterjee, S., Sarddar, D., Saha, J., Banerjee, S., Mondal, A. & Naskar, M.K. 2012. An improved mobility management technique for IEEE 802.11 based WLAN by predicting the direction of the mobile node. *2012 National Conference on Computing and Communication Systems, NCCCS 2012 – Proceeding,* pp. 242–246.

Chávez-Santiago, R., Szydelko, M., Kliks, A., Foukalas, F., Haddad, Y., Nolan, K.E., Kelly, M.Y., Masonta, M.T. & Balasingham, I. 2015. 5G: The convergence of wireless communications. *Wireless Personal Communications,* 83(3), 1617–1642.

Chen, J., Yang, C. & Mai, Y. 2015a. A novel smart forwarding scheme in LTE- Advanced networks. *China Communications,* 12(3), 120–131.

Chen, S., Sun, S., Wang, Y., Xiao, G. & Tamrakar, R. 2015b. A comprehensive survey of TDD-based mobile communication systems from TD-SCDMA 3G to TD-LTE(A) 4G and 5G directions. *China Communications,* 12(2), 40–60.

Chen, X., Kim, K., Lee, B. & Youn, H. 2014a. DIHAT: Differential integrator handover algorithm with TTT window for LTE-based systems. *EURASIP Journal on Wireless Communications and Networking,* 2014(1), 162.

Chen, X., Kim, M.J., Yoo, S.H., Park, N.Y. & Youn, H.Y. 2014b. Efficient and prompt handover in LTE-based systems by predicting the target eNodeBs. *2014 International Conference on Cyber-Enabled Distributed Computing and Knowledge Discovery,* pp. 406–413.

Cheng-Chung, L. 2013. *Handover mechanisms in 3GPP long term evolution (LTE).* PhD Thesis, University of Technology, Sydeny. https://opus.lib.uts.edu.au/ handle/10453/ 28011.

Cheng-Chung, L., Sandrasegaran, K., Xinning, Z. & Zhuliang, X. 2012. On the performance of capacity integrated CoMP handover algorithm in LTE- Advanced. *2012 18th Asia-Pacific Conference on Communications (APCC)*, pp. 871–876. IEEE.

Chukhno, N., Orsino, A., Torsner, J., Iera, A. & Araniti, G. 2023. 5G NR sidelink multi-hop transmission in public safety and factory automation scenarios. *IEEE Network*, 37, 129–136.

Cisco. 2017. Cisco Visual Networking Index: Global Mobile Data Traffic Forecast Update, 2015–2020. White Paper. wwwcisco.com/c/en/us/solutions/collateral/service-provider/visual-networking-index-vni/mobile-white-paper-c11-520862.html [03-04-2017].

Dahlman, E., Parkvall, S., Sköld, J. & Beming, P. 2007. *3G Evolution HSPA and LTE for Mobile Broadband*. 1.0 ed. Elsevier.

Damaševičius, R., Bacanin, N. & Misra, S. 2023. From sensors to safety: Internet of Emergency Services (IoES) for emergency response and disaster management. *Journal of Sensor and Actuator Networks*, 12, 41.

Damnjanovic, A., Montojo, J., Wei, Y., Ji, T., Luo, T., Vajapeyam, M., Yoo, T., Song, O. & Malladi, D. 2011. A survey on 3GPP heterogeneous networks. *IEEE Wireless Communications,* 18(3), 10–21.

Davaasambuu, B., Yu, K. & Sato, T. 2015. Self-optimization of handover parameters for long-term evolution with dual wireless mobile relay nodes. *Future Internet,* 7(2), 196–213.

Deswal, S. & Singhrova, A. 2016. Handover algorithm for heterogeneous networks. *2016 International Conference on Computing for Sustainable Global Development,* pp. 3358–3364.

Fall, K. & Varadhan, K. 2011. Ns notes and documents. The VINT Project, UC Berkeley, LBL, USC/ISI, and Xerox PARC.

Fazio, M., Paone, M., Puliafito, A. & Villari, M. Heterogeneous sensors become homogeneous things in smart cities. *2012 Sixth International Conference on Innovative Mobile and Internet Services in Ubiquitous Computing*, 2012. IEEE, 775–780.

Fei, M. & Fan, P. 2012. Position-assisted fast handover schemes for LTE-advanced network under high mobility scenarios. *Journal of Modern Transportation,* 20(4), 268–273.

Ge, H., Wen, X., Zheng, W., Lu, Z. & Wang, B. 2009. A history-based handover prediction for LTE systems. *Proceedings – 1st International Symposium on Computer Network and Multimedia Technology, CNMT 2009,* pp. 1–4.

Ghanem, K., Alradwan, H., Motermawy, A. & Ahmad, A. 2012. Reducing ping-pong handover effects in intra EUTRA networks. *Proceedings of the 2012 8th International Symposium on Communication Systems, Networks and Digital Signal Processing, CSNDSP 2012,* pp. 1–5.

Ghosh, A., Ratasuk, R., Mondal, B., Mangalvedhe, N. & Thomas, T. 2010. LTE-advanced: Next-generation wireless broadband technology [Invited Paper]. *IEEE Wireless Communications,* 17(3), 10–22.

Gódor, G., Jakó, Z., Knapp, Á. & Imre, S. 2015. A survey of handover management in LTE-based multi-tier femtocell networks: Requirements, challenges and solutions. *Computer Networks,* 76(November), 17–41.

Goldwnamith, A. 2005. *Wireless communications*. 1.0 ed. Cambridge University Press.

Graziosi, F. & Santucci, F. 2002. A general correlation model for shadow fading in mobile radio systems. *IEEE Communications Letters,* 6(3), 102–104.

Habibzadeh, A., Moghaddam, S.S., Razavizadeh, S.M. & Shirvanimoghaddam, M. 2015. A novel handover decision-making algorithm for HetNets. *2015 IEEE International Symposium on Signal Processing and Information Technology (ISSPIT)*, pp. 438–442.

Hakak, S., Gadekallu, T. R., Maddikunta, P. K. R., Ramu, S. P., ParimalA, M., DE Alwis, C. & Liyanage, M. 2023. Autonomous vehicles in 5G and beyond: A survey. *Vehicular Communications*, 39, 100551.

Han, J. & Wu, B. 2010. Handover in the 3GPP long term evolution (LTE) systems. *2010 Global Mobile Congress, GMC'2010* .

Hansen, R., Wind, R., Jensen, C.S. & Thomsen, B. 2009. Seamless indoor/outdoor positioning handover for location-based services in streamspin. *Proceedings – IEEE International Conference on Mobile Data Management,* pp. 267–272.

Hätty, B., Dreßler, H.J., Kröner, H., Schnabl, G., Schopp, M. & Splett, A. 2016. Adaptive techniques in advanced 4G cellular wireless networks. *EURASIP Journal on Advances in Signal Processing,* 2016(1), 19.

Huang, Y.S., Leu, F.Y., Liu, J.C., Huang, Y.L. & Chu, W.C.C. 2013. A handover scheme for LTE wireless networks under the assistance of GPS. *Proceedings – 2013 8th International Conference on Broadband, Wireless Computing, Communication and Applications, BWCCA 2013,* pp. 399–403.

Iftikhar, S., Asim, M., Zhang, Z., MuthannA, A., Chen, J., EL-Affendi, M., Sedik, A. & ABD EL-Latif, A. A. 2023. Target detection and recognition for traffic congestion in smart cities using deep learning-enabled UAVs: A review and analysis. *Applied Sciences,* 13, 3995.

Isa, I.N.M., Dani Baba, M., Yusof, A.L. & Rahman, R.A. 2015. Handover parameter optimization for self-organizing LTE networks. *2015 IEEE Symposium on Computer Applications & Industrial Electronics (ISCAIE),* pp. 1–6. September.

ITU-R. 2008. Requirements related to technical performance for IMT-Advanced radio interface (s). www.itu.int/dms_pub/itu-r/opb/rep/R-REP-M.2134-2008-PDF-E.pdf [03-06-2017].

Jansen, T., Balan, I., Turk, J., Moerman, I. & Kurner, T. 2010. Handover parameter optimization in LTE self-organizing networks. *2010 IEEE 72nd Vehicular Technology Conference – Fall,* pp. 1–5. September.

Jeong, B., Shin, S., Jang, I., Sung, N.W. & Yoon, H. 2011. A smart handover decision algorithm using location prediction for hierarchical macro/femto- cell networks. *2011 IEEE Vehicular Technology Conference (VTC Fall),* pp. 1–5.

Jimaa, S. & Alfadhl, Y. 2011. LTE-A an overview and future research areas. *2011 IEEE 7th International Conference on Wireless and Mobile Computing, Networking and Communications (WiMob),* pp. 395–399.

Kanchi, S., Sandilya, S., Bhosale, D., Pitkar, A. & Gondhalekar, M. 2013. Overview of LTE-A technology. *2013 IEEE Global High Tech Congress on Electronics,* volume 3, pp. 195–200.

Khattab, A.K.F. & Elsayed, K.M.F. 2006. Opportunistic scheduling of delay sensitive traffic in OFDMA-based wireless networks. *Proceedings – WoWMoM 2006: 2006 International Symposium on a World of Wireless, Mobile and Multimedia Networks* 2006, pp. 279–288.

Khattab, O., Khan, M. & Alothman, B. 2023. Designing a vertical handover algorithm for security-constrained applications. *Applied Sciences,* 13, 2166.

Kim, J., Lee, G. & In, H.P. 2014. Adaptive time-to-trigger scheme for optimizing LTE handover. *International Journal of Control and Automation,* 7(4), 35–44.

Kim, R., Jung, I., Yang, X. & Chou, C.C. 2010. Advanced handover schemes in IMT- advanced systems [WiMAX/LTE Update]. *IEEE Communications Magazine,* 48(8), 78–85.

Kurjenniemi, J. & Henttonen, T. 2008. Effect of measurement bandwidth to the accuracy of inter-frequency RSRP measurements in LTE. *IEEE International Symposium on Personal, Indoor and Mobile Radio Communications, PIMRC.*

Lee, B.G. & Choi, S. 2008. *Broadband wireless access and local networks: Mobile WiMAX and WiFi.* Norwood: Artech House.

Lee, D.W., Gil, G.T. & Kim, D.H. 2010. A cost-based adaptive handover hysteresis scheme to minimize the handover failure rate in 3GPP LTE system. *EURASIP Journal on Wireless Communications and Networking,* 2010(1), 750173.

Lee, J., Kim, Y., Lee, H., Ng, B., Mazzarese, D., Liu, J., Xiao, W. & Zhou, Y. 2012. Coordinated multipoint transmission and reception in LTE-advanced systems. *IEEE Communications Magazine,* 50(11), 44–50.

Lee, J. C., Cho, S.-P. & Kim, H. 2005. Position based handover control method. *Computational Science and Its Applications – ICCSA 2005,* Pt 2, 781–788.

Lei, P.R., Shen, T.J., Peng, W.C. & Su, I.J. 2011. Exploring spatial-temporal trajectory model for location prediction. *Proceedings – IEEE International Conference on Mobile Data Management,* 1, 58–67.

Li, X.W. & Wang, J. 2013. The optimized method of reducing unnecessary handover in LTE system. *2013 Third International Conference on Instrumentation, Measurement, Computer, Communication and Control*, pp. 1224–1227.

Li, Y., Cao, B. & Wang, C. 2016. Handover schemes in heterogeneous LTE networks: Challenges and opportunities. *IEEE Wireless Communications* 23(2), 112–117.

Lin, P.C., Casanova, L.F.G. & Fatty, B.K.S. 2016. Data-driven handover optimization in next generation mobile communication networks. *Mobile Information Systems,* 2016, 1–11.

Lin, Y.B., Pang, A.C. & Chung-Hwa Rao, H. 2005. Impact of mobility on mobile telecommunications networks. *Wireless Communications and Mobile Computing,* 5(7), 713–732.

Liu, C., Wei, J., Huang, S. & Cao, Y. 2012. A distance-based handover scheme for femtocell and macrocell overlaid networks. *2012 8th International Conference on Wireless Communications, Networking and Mobile Computing*, pp. 1–4.

Lopez-Perez, D., Guvenc, I., de la Roche, G., Kountouris, M., Quek, T. & Zhang, J. 2011. Enhanced intercell interference coordination challenges in heterogeneous networks. *IEEE Wireless Communications,* 18(3), 22–30.

Lu, X., Qu, Z., Lio, P., Hui, P., Li, Q., Lu, P. & Bie, R. 2014. Directional communication with movement prediction in mobile wireless sensor networks. *Personal and Ubiquitous Computing,* 18(8), 1941–1953.

Maheswari, A., Prithiviraj, A. & Krishnamoorthy, K. 2016. Markov based VHO to improve the Handover performance among heterogeneous wireless networks in PMIPv6 domain. *2016 International Conference on Circuit, Power and Computing Technologies (ICCPCT)*, pp. 1–6.

Márquez-Barja, J., Calafate, C.T., Cano, J.C. & Manzoni, P. 2011. An overview of vertical handover techniques: Algorithms, protocols and tools. *Computer Communications,* 34(8), 985–997.

Marshoud, H., Otrok, H., Barada, H., Estrada, R., Jarray, A. & Dziong, Z. 2016. Realistic framework for resource allocation in macro–femtocell networks based on genetic algorithm. *Telecommunication Systems,* 63(1), 99–110.

Mishra, P. & Singh, G. 2023. Energy management systems in sustainable smart cities based on the internet of energy: A technical review. *Energies*, 16, 6903.

Miyim, A.M., Ismail, M. & Nordin, R. 2014. Vertical handover solutions over LTE- advanced wireless networks: An overview. *Wireless Personal Communications,* 77(4), 3051–3079.

Moon, J.M. & Cho, D.H. 2010. Novel handoff decision algorithm in hierarchical macro/femtocell networks. *2010 IEEE Wireless Communication and Networking Conference*, pp. 1–6.

Munoz, P., Barco, R. & De La Bandera, I. 2013. On the potential of handover parameter optimization for self-organizing networks. *IEEE Transactions on Vehicular Technology,* 62(5), 1895–1905.

Naeem, B., Ngah, R., & Mohd Hashim, S.Z. 2016. Handovers in small cell based heterogeneous networks. *2016 International Conference on Computing, Electronic and Electrical Engineering (ICE Cube),* pp. 0–3.

NS-3 Consortium. (2017). *ns-3 Network Simulator*. www.nsnam.org/ [11-04-2017].

Olaverri-MonreaL, C. 2016. Intelligent technologies for mobility in smart cities. *Hiradastechnika Journal*, 71, 29–34.

Patzold, M., Killat, U. & Laue, F. 1996. A deterministic digital simulation model for Suzuki processes with application to a shadowed Rayleigh land mobile radio channel. *IEEE Transactions on Vehicular Technology,* 45(2), 318–331.

Peralta-Ochoa, A.M., Chaca-Asmal, P.A., Guerrero-Vásquez, L.F., Ordoñez-Ordoñez, J.O. & Coronel-González, E.J. 2023. Smart healthcare applications over 5G networks: A systematic review. *Applied Sciences*, 13, 1469.

Piro, G., Grieco, L.A., Boggia, G., Capozzi, F. & Camarda, P. 2011. Simulating LTE cellular systems: An open source framework. *IEEE Transactions on Vehicular Technology,* 60(2), 498–513.

Pons, M., Valenzuela, E., Rodríguez, B., Nolazco-Flores, J.A. & Del-Valle-Soto, C. 2023. Utilization of 5G technologies in IoT applications: Current limitations by interference and network optimization difficulties—A review. *Sensors*, 23, 3876.

Rafique, W., Barai, J., Fapojuwo, A.O. & Krishnamurthy, D. 2024. A survey on beyond 5G network slicing for smart cities applications. *IEEE Communications Surveys & Tutorials*, 26(2), 1342–1372. https://doi.org/10.1109/COMST.2024.3387576

Rani, P. & Sharma, R. 2024. Intelligent transportation system performance analysis of indoor and outdoor internet of vehicle (IOV) applications towards 5G. *Tsinghua Science and Technology*, 29, 1785–1795.

Ramachandran, R. 2003. Evolution to 3G mobile communication. *Resonance,* 8(11), 37–51.

Salman, H.A., Ibrahim, L.F. & Fayed, Z. 2014. Overview of LTE-advanced mobile network plan layout. *2014 5th International Conference on Intelligent Systems, Modelling and Simulation*, pp. 585–590.

Sánchez, J.J., Gómez, G., Morales-Jiménez, D. & Entrambasaguas, J.T. 2006. Performance evaluation of OFDMA wireless systems using WM-SIM platform. *Proceedings of the International Workshop on Mobility Management and Wireless Access – MobiWac '06,* p. 131.

Scalable Network Technologies. QualNet Network Simulator Software. www.scalable-networks.com/qualnet-network-simulator-software [11-02-2017].

Sesia, S., Toufik, I. & Baker, M. 2011. *LTE: The UMTS long term evolution from theory to practice.* 2.0 ed. Chichester, UK: John Wiley & Sons.

Shayea, I., Ismail, M. & Nordin, R. 2012. Advanced handover techniques in LTE-Advanced system. *2012 International Conference on Computer and Communication Engineering, ICCCE 2012* (July), pp. 74–79.

Shen, Y., Luo, T. & Win, M.Z. 2012a. Neighboring cell search for LTE systems. *IEEE Transactions on Wireless Communications,* 11(3), 908–919.

Shen, Z., Papasakellariou, A., Montojo, J., Gerstenberger, D. & Xu, F. 2012b. Overview of 3GPP LTE-advanced carrier aggregation for 4G wireless communications. *IEEE Communications Magazine,* 50(2), 122–130.

Sipila, K., Jasberg, M., Laiho-Steffens, J. & Wacker, A. 1999. Soft handover gains in a fast power controlled WCDMA uplink. *1999 IEEE 49th Vehicular Technology Conference (Cat. No.99CH36363),* 2, 1594–1598.

Suleiman, K.E., Taha, A.E.M. & Hassanein, H.S. 2016. Handover-related self- optimization in femtocells: A survey and an interaction study. *Computer Communications,* 73, 82–98.

Sung, N.W., Pham, N.T., Yoon, H., Lee, S. & Hwang, W.J. 2013. Base station association schemes to reduce unnecessary handovers using location awareness in femtocell networks. *Wireless Networks,* 19(5), 741–753.

Tabia, N., Gondran, A., Baala, O. & Caminada, A. 2012. Interference model and antenna parameters setting effects on 4G-LTE networks coverage. *Proceedings of the 7th ACM Workshop on Performance Monitoring and Measurement of Heterogeneous Wireless and Wired Networks – PM2HW2N '12,* p. 175.

Tao, M., Yuan, H., Hong, X. & Zhang, J. 2016. SmartHO: Mobility pattern recognition assisted intelligent handoff in wireless overlay networks. *Soft Computing,* 20(10), 4121–4130.

Tran, T.T., Shin, Y. & Shin, O.S. 2012. Overview of enabling technologies for 3GPP LTE-advanced. *EURASIP Journal on Wireless Communications and Networking,* 2012(1), 54.

Ulvan, A., Bestak, R. & Ulvan, M. 2011. Handover procedure and decision strategy in LTE-based femtocell network. *Telecommunication Systems,* 52(4), 2733–2748.

Venkata Rajesh, M., Jayarekha, P., Masali, S. & Bhargavi, D.K. 2015. Effective algorithm to minimize target FAPs during hand-in and hand-off for 3GPP-LTE femtocell network. *2015 IEEE International Advance Computing Conference (IACC),* pp. 211–216.

Venkatesan, V.K., Izonin, I., Periyasamy, J., Indirajithu, A., Batyuk, A. & Ramakrishna, M.T. 2022. Incorporation of energy efficient computational strategies for clustering and routing in heterogeneous networks of smart city. *Energies*, 15, 7524.

Volkan, Y., Ulas, K. & Oguz, S. 2014. A new control plane for 5G network architecture with a case study on unified handoff, mobility, and routing management. *IEEE Communications Magazine,* 52(11), 76–85.

Wanalertlak, W., Lee, B., Yu, C., Kim, M., Park, S.M. & Kim, W.T. 2011. Behavior- based mobility prediction for seamless handoffs in mobile wireless networks. *Wireless Networks,* 17(3), 645–658.

Wang, H.L., Kao, S.J., Hsiao, C.Y. & Chang, F.M. 2014a. A moving direction prediction-assisted handover scheme in LTE networks. *EURASIP Journal on Wireless Communications and Networking,* 2014(1), 190.

Wang, T., Luo, H., Zeng, X., YU, Z., Liu, A. & Sangaiah, A. K. 2020. Mobility based trust evaluation for heterogeneous electric vehicles network in smart cities. *IEEE Transactions on Intelligent Transportation Systems*, 22, 1797–1806.

Wang, Y.H., Huang, G.R. & Tung, Y.C. 2014b. A handover prediction mechanism based on LTE-A UE history information. *2014 International Conference on Computer, Information and Telecommunication Systems, CITS 2014.*

Watanabe, Y., Matsunaga, Y., Kobayashi, K., Sugahara, H. & Hamabe, K. 2011. Dynamic neighbor cell list management for handover optimization in LTE. *IEEE Vehicular Technology Conference.*

Wu, C.S., Chu, Y.S. & Fang, C.H. 2013. The periodic scan and velocity decision hand-over scheme for next generation femtocell/macrocell overlay networks. *International Conference on ICT Convergence,* pp. 201–206.

Xenakis, D., Passas, N., Merakos, L. & Verikoukis, C. 2014. Mobility management for femtocells in LTE-advanced: Key aspects and survey of handover decision algorithms. *IEEE Communications Surveys and Tutorials,* 16(1), 64–91.

Xenakis, D., Passas, N., Merakos, L. & Verikoukis, C. 2016. Handover decision for small cells: Algorithms, lessons learned and simulation study. *Computer Networks,* 100, 64–74.

Xu, P., Fang, X., He, R. & Xiang, Z. 2013. An efficient handoff algorithm based on received signal strength and wireless transmission loss in hierarchical cell networks. *Telecommunication Systems,* 52(1), 317–325.

Yang, F., Deng, H., Jiang, F. & Deng, X. 2015. Handover optimization algorithm in LTE high-speed railway environment. *Wireless Personal Communications,* 84(2), 1577–1589.

Yang, S.C. & Winter, P. 2015. LTE-Advanced and IEEE 802.11ac. *International Journal of Information and Learning Technology,* 32(4), 221–234.

Yilmaz, O.N.C., Hämäläinen, J. & Hämäläinen, S. 2013. Optimization of adaptive antenna system parameters in self-organizing LTE networks. *Wireless Networks,* 19(6), 1251–1267.

Zhang, H., Wen, X., Wang, B., Zheng, W. & Sun, Y. 2010. A novel handover mechanism between femtocell and macrocell for LTE based networks. *2010 Second International Conference on Communication Software and Networks*, pp. 228–231.

Zhou, Y. & Ai, B. 2014. Handover schemes and algorithms of high-speed mobile environment: A survey. *Computer Communications,* 47(April), 1–15.

Zhu, H. & Kwak, K.S. 2007. Performance analysis of an adaptive handoff algorithm based on distance information. *Computer Communications,* 30(6), 1278–1288.

Appendices

APPENDIX A LTE-SIM SIMULATOR

This work has been completed using a dynamic framework simulator called LTE-Sim, an open-source structure to simulate LTE-A systems. In order to guarantee flexibility, modularity, adaptability, and performance, LTE-Sim has been transcribed into C++, using the object-oriented prototype, as an event-driven simulator system. In addition, LTE-Sim envelops numerous parts of the LTE-A system, including both E-UTRAN and EPC. It bolsters heterogeneous and homogenous multi-cell environments, multi-client environment, client mobility, frequency reuse techniques, handover procedure, and QoS management. Four categories of network nodes are demonstrated: UE, eNB, HeNB, and MME/GW. Four types of data traffic generators at the application layer have been applied and the management of data radio bearer has been maintained. Moreover, the recognized scheduling approaches such as Modified Largest Weighted Delay First, Exponential Proportional Fair, Proportional Fair, and Log and Exp rules, Channel Quality Indicator feedback, frequency reuse techniques, AMC scheme, and models of physical layer have been established. In the LTE-Sim, there are four main components: the simulator, network manager, flow manager, and frame manager. These components are illustrated in Table A.1.

In the handover procedure, the LTE-Sim performs inter-cell handover in order to support user mobility. Three types of mobility prototypes are maintained: Manhattan, random direction, and random walk. For each of them, a single class was established, and in each mobility prototype class, the m_speed and m_speed direction variables are utilized to characterize the speed and mobile direction of the client, respectively. At the point where the random direction model is used, the UE randomly selects the speed direction that remains constant over time, and moves toward the border region. Once the simulation border region is obtained, the UE selects another speed direction. The travel distance is equivalent to 200, 400, and 1,000 m by default. At the point where UE speed is equivalent to 3, 25, 30, 120, and 150 km/h, respectively, the UE mobility is accomplished by the network manager, as each TTI refreshes the UE position as per the selected mobility model and parameters, and validates through the NetworkManager::HandOverProcedure() function if the handover technique is needed. In LTE-Sim both cell re-selection and handover strategies are achieved.

TABLE A.1
The main components of LTE-Sim

Component	Functionalities	Important Method	Method Description
Simulator	1. Creates/handles/ ends an events	Schudle()	Creates a new event and inserts it into the calendar
		RunOneEvent()	Executes an event
		Run()/Stop()	Starts/ends the simulation
Network manager	1. Creates devices 2. Handles UE positions 3. Manages the handover 4. Implements frequency reuse techniques	CreatUserEquipment() CreateCell() UpdateUserPosition() HandOverProcedure() RUNFrequencyReuse() CreateApplication 0	Creates a UE device Creates LTE-A cell Updates UE position Handles the handover procedure Implements frequency reuse techniques
Flow manager	1. Handles applications		Creates an application
Frame manager	1. Defines LTE-A frame structure 2. Schedules frames and subframes	StartFrame() and StopFram() StartSubFrame() and StopSubFrame()	Handles the start and the end of the LTE-A frame Handles the start and the end of the LTE-A sub frame

In addition, handover decisions are executed through the handover manager procedure, which is defined for every UE. Handover management involves the following steps in LTE-Sim:

1. For the UE that triggers the handover procedure, the functionHandOverMa nager::SelectTargetENodeB () is utilized for choosing a new target cell by applying Position-based or Power-based algorithms.
2. All information about the UE is transferred from the old cell to the new one.
3. New radio bearer is produced among the UE and the new target eNB.
4. The UE updates the list of the available subchannels for downlink and uplink, according to those assigned to the new target cell.

Currently, LTE-Sim supports handover decision algorithm in two models, distance and power (RSRP). In the distance model, the handover is triggered when the distance between the UE and the target cell is less than the distance between the UE and its serving cell. On the contrary, in the power model, the handover decision is triggered based on the RSRP. In this work to evaluate the performance of the LTE-A handover, both methods are used. Furthermore, since the LTE-Sim does not support the admission controller in handover procedure, some alterations have been done on the handover procedure in this work, such as admission controller, handover among HeNBs, tracing UE positions and directions, and the UE speed becomes 25 km/h. To implement the LTE-Sim simulator, hardware and software requirements are needed. The simulation was run on HP laptop with 350 GB, 6 GB RAM, 2.4 GHZ, 3 MB Intel Core i3 processor. Moreover, the Linux-Centos Operating System, LTE-Sim version R5, and Matlab version 2013a were used to implement the proposed algorithms and draw the simulation topologies.

APPENDIX B TOTAL COMPARISON RESULTS OF PERFORMANCE METRICS BETWEEN MDD-VHD AND REFERENCE

TABLE B.1
Results of average handovers number and average signaling measurement number between MDD-VHD and Deswal and Singhrova (2016)

Performance Metric	UEs number	MDD-VHD				Deswal & Singhrova (2016)			
		HeNBs number				HeNBs number			
		20	40	60	80	20	40	60	80
Average handover number	10	0.0307	0.0923	0.123	0.161	0.061	0.138	0.315	0.338
	20	0.0923	0.169	0.238	0.407	0.130	0.369	0.484	0.676
	40	0.138	0.276	0.538	0.684	0.323	0.969	1.123	1.5
	80	0.415	0.823	1.084	1.4	0.784	1.184	1.923	2.892
Average of signaling measurement number	10	14,489	19,230	18,140	18,577	15,384 500.59	30,768 998.82	46,153 511.24	61,538 004.14
	20	20,624	29,326	38,326	42,326	30,768 998.22	61,538 006.51	92,306 999.41	12,307 6003
	40	77,249	83,608	106,357	105,924	61,537 996.45	12,307 6013	18,461 3998.8	246,152 005.9
	80	140,121	142,983	170,912	165,687	12,307 5992.9	24,615 2026	36,922 7997.6	49,230 4011.8

TABLE B.2
Results of packetdelay, packet loss, and throughput between MDD-VHD and Deswal and Singhrova (2016) in CBR application

		MDD-VHD				Deswal & Singhrova (2016)			
		Number of HeNBs				Number of HeNBs			
Performance Metric	UEs umber	20	40	60	80	20	40	60	80
Packet delay ratio	**10**	0.00219	0.00283	0.00325	0.00533	0.00379	0.10403	0.1978	0.53091
	20	0.00223	0.00271	0.00485	0.01292	0.0074	0.13829	0.49186	0.93246
	40	0.00248	0.00388	0.00845	0.01476	0.01244	0.21247	0.54841	1.01028
	80	0.00882	0.0519	0.06625	0.07569	0.025	0.36286	1.01094	1.24456
Packet loss ratio	**10**	0.00453	0.0167	0.02564	0.05269	0.04399	0.14052	0.1626	0.19989
	20	0.00698	0.02501	0.03861	0.05856	0.07484	0.14983	0.20788	0.2187
	40	0.02313	0.05488	0.07036	0.07489	0.08228	0.22606	0.24685	0.263
	80	0.17802	0.19847	0.28	0.28297	0.19631	0.32	0.32084	0.34455
Throughput	**10**	503.86024 62	492.43852 31	490.9432	480.50775 38	488.11163 08	433.35735 38	401.28744 62	395.52886 15
	20	1005.1752 62	994.48529 23	976.82775 38	953.92067 69	936.45403 08	862.22873 85	847.78455 38	813.01033 85
	40	1978.8214 77	1916.5905 85	1872.2081 23	1884.9660 92	1864.1588 31	1487.5283 08	1566.3350 15	1528.0611 08
	80	3331.7388 92	3252.0731 69	2917.8525 54	2900.8949 54	3263.2085 54	2751.9035 08	2754.9259 69	2661.6750 77

TABLE B.3
Results of packet delay, packet loss, and throughput between MDD-VHD and Deswal and Singhrova (2016) using VoIP application

		MDD-VHD				Deswal and Singhrova (2016)			
		Number of HeNBs				Number of HeNBs			
Performance Metric	UEs umber	20	40	60	80	20	40	60	80
Packet delay ratio	**10**	0.00147	0.0019	0.01311	0.03133	0.06011	1.30716	1.50758	1.77601
	20	0.00222	0.00796	0.01906	0.0511	0.07952	1.71227	1.77808	2.11548
	40	0.01079	0.02539	0.03179	0.0633	1.42863	2.27966	2.38335	2.89877
	80	0.01212	0.03587	0.04463	0.07455	2.49934	2.52638	2.72229	3.25781
Packet loss ratio	**10**	0.00034	0.00211	0.00246	0.00582	0.00806	0.01331	0.02122	0.0248
	20	0.00152	0.00329	0.00511	0.00759	0.02157	0.02232	0.02269	0.03449
	40	0.00158	0.00537	0.00545	0.00683	0.00231	0.02386	0.02782	0.03693
	80	0.00157	0.00522	0.01088	0.01753	0.03931	0.04763	0.05368	0.05732
Throughput	**10**	92.68443077	92.4112	92.24270769	91.43667692	91.96947692	91.48221538	90.74904615	90.41661538
	20	1005.175262	994.4852923	976.8277538	953.9206769	181.4320615	181.2248615	181.2931692	179.0367385
	40	1978.821477	1916.590585	1872.208123	1884.966092	370.0068308	362.0171077	360.5484923	357.1695385
	80	3331.738892	3252.073169	2917.852554	2900.894954	712.5767385	706.404	701.9161846	699.2180308

APPENDIX C TOTAL COMPARISON RESULTS OF PERFORMANCE METRICS IN MDD-VHD IN THREE DIFFERENT SCENARIOS OF HENBS DISTRIBUTION

TABLE C.1
Results of average handovers number and average signaling measurement number in MDD-VHD in three different scenariois of HeNBs

		MDD-VHD							
		Close Number of HeNBs				Middle Number of HeNBs			
Performance Metric	UEs umber	20	40	60	80	20	40	60	80
Average handover number	10	0.1	0.161	0.261	0.315	0.061	0.138	0.146	0.261
	20	0.169	0.369	0.446	0.523	0.138	0.176	0.269	0.4
	40	0.592	0.769	1.130	1.215	0.215	0.407	0.730	0.807
	80	0.853	1.615	2.092	2.446	0.530	1.007	1.223	1.615
Average of signaling measurement number	10	133.4	85.7	229.6	225.2	51.0	154.2	171.0	180.2
	20	278.3	299.6	351.7	222.3	287.6	324.8	329.7	332.8
	40	318.7	379.0	452.1	481.9	691.9	567.8	492.9	918.5
	80	1246.4	1414.1	1647.1	1246.6	1196.5	1130.6	932.6	1427.1

TABLE C.2
Results of packet delay, packet loss, and throughput using CBR application in MDD-VHD in three different scenarios of HeNBs distribution

		MDD-VHD							
Performance		**Number of HeNBs**				**Number of HeNBs**			
Metric	**UEs umber**	**20**	**40**	**60**	**80**	**20**	**40**	**60**	**80**
Packet delay ratio	**10**	0.00491	0.00516	0.00935	0.01523	0.00452	0.00464	0.00781	0.00905
	20	0.00671	0.01077	0.01524	0.01706	0.00454	0.00462	0.00785	0.02142
	40	0.01384	0.02021	0.02712	0.03703	0.00223	0.01081	0.02246	0.02784
	80	0.01336	0.03631	0.03727	0.0424	0.00667	0.01199	0.01454	0.02399
Packet loss ratio	**10**	0.00677	0.02697	0.0274	0.04236	0.00111	0.01884	0.03688	0.03861
	20	0.16499	0.21432	0.22281	0.24055	0.06826	0.10505	0.11296	0.12337
	40	0.49435	0.50071	0.52821	0.53144	0.20393	0.27592	0.29878	0.31861
	80	0.62234	0.67238	0.67309	0.68218	0.20393	0.48116	0.48808	0.50764
Throughput	**10**	513.18	502.74	502.524	494.79	516.10	506.94	497.62	496.73
		21538	67077		28615	91692	63385	44308	36
	20	862.86	811.89	803.11	784.79	962.82	924.80	916.63	905.87
		50462	68	57538	00923	89846	96	30462	94462
	40	1045.03	969.89	954.39	947.71	1645.26	1496.46	1449.22	1408.24
		9938	2	79077	66769	8985	8431	2585	4369
	80	1561.02	1354.19	1351.26	1313.68	1645.26	2144.61	2116.00	2035.13
		1846	0031	3015	9046	8985	1446	9415	4708

Continue … .

TABLE C.3
Results of packet delay, packet loss, and throughput using VoIP application in MDD-VHD in three different scenarios of HeNBs distribution

		MDD-VHD							
		Number of HeNBs				Number of HeNBs			
Performance Metric	**UEs umber**	**20**	**40**	**60**	**80**	**20**	**40**	**60**	**80**
Packet delay ratio	**10**	0.00153	0.00167	0.06424	0.10706	0.00172	0.0564	0.03232	0.07077
	20	0.01463	0.04188	0.09313	0.2113	0.002	0.01027	0.0199	0.03176
	40	0.00629	0.03108	0.05383	0.09504	0.00223	0.01081	0.02246	0.02784
	80	0.01411	0.04867	0.07884	0.08884	0.00421	0.01406	0.02501	0.03535
Packet loss ratio	**10**	0.0015	0.00189	0.00273	0.00589	0.00032	0.00037	0.00204	0.00474
	20	0.00157	0.00196	0.00384	0.00622	0.00088	0.00091	0.00371	0.00502
	40	0.0015	0.00276	0.00487	0.00686	0.00104	0.0025	0.00404	0.00635
	80	0.003	0.00328	0.00586	0.00694	0.0016	0.00276	0.00569	0.0091
Throughput	**10**	92.56	92.53	92.46	92.15	92.68	92.68	92.43	92.27
		830769	415385	356923	846154	670769	215385	852308	686154
	20	185.18	184.82	184.68	184.15	185.23	185.08	184.73	184.33
		89846	46769	35077	98154	90769	19692	81538	96923
	40	370.30	369.84	369.05	368.29	370.53	370.10	369.43	368.55
		73846	06154	96308	23077	73538	92923	98769	87077
	80	740.82	739.26	737.51	737.01	741.09	740.24	739.87	738.01
		42462	91077	36	95077	06462	59077	24923	95077

Continue… ..

Continue … .. MDD-VHD					
Performance Metric	UEs umber	Edge Number of HeNBs			
		20	40	60	80
Average handover number	**10**	0.015	0.053	0.138	0.161
	20	0.107	0.130	0.223	0.307
	40	0.161	0.315	0.492	0.507
	80	0.353	0.592	0.784	1,076
Average of signaling measurement number	**10**	134.5307692	147.9230769	139.5384615	182.4538462
	20	325.5846154	352.0153846	112.4923077	228.0153846
	40	594.2230769	643.1384615	818.1307692	1314.707692
	80	1077.853846	1099.869231	814.8	1274.515385

Continue…… MDD-VHD					
Performance Metric	UEs umber	Edge Number of HeNBs			
		20	40	60	80
Packet delay ratio	**10**	0.01895	0.01393	0.02291	0.04141
	20	0.01549	0.0172	0.03321	0.11707
	40	0.04508	0.03648	0.12052	0.20836
	80	0.06225	0.06732	0.12052	0.18648
Packet loss ratio	**10**	0.01137	0.01887	0.02412	0.05527
	20	0.21887	0.28032	0.30677	0.31400
	40	0.52224	0.53348	0.54205	0.54305
	80	0.65228	0.67464	0.69948	0.70125
Throughput	**10**	500.9014154	497.9107602	405.5564308	477.4852023
	20	807.1881231	734.3045231	716.3552	707.8604923
	40	987.3004615	964.1652308	944.5087692	942.5307692
	80	1437.201815	1344.868123	1242.190877	1234.850523

Continue...... MDD-VHD					
Performance Metric	**UEs umber**	**Edge Number of HeNBs**			
		20	**40**	**60**	**80**
Packet delay ratio	**10**	0.06432	0.08344	0.1485	0.25232
	20	0.05594	0.09924	0.15871	0.2741
	40	0.07574	0.11091	0.20009	0.28764
	80	0.04037	0.13539	0.28764	0.29986
Packet loss ratio	**10**	0.00131	0.00788	0.01057	0.01513
	20	0.00416	0.00867	0.01141	0.0165
	40	0.00467	0.01129	0.01304	0.02315
	80	0.0112	0.01287	0.01754	0.03337
Throughput	**10**	91.67803077	91.5232	91.31372308	89.62196923
	20	185.1411692	183.8251077	183.3173538	182.3724308
	40	369.3214769	367.9416615	366.9443692	366.0290462
	80	738.2695385	733.3559385	728.7201231	724.5601846

Index

Note: Page locators in **bold** and *italics* represents tables and figures, respectively.

H

N

O

P

Q

R

S

T

U

V

W

X

For Product Safety Concerns and Information please contact our EU
representative GPSR@taylorandfrancis.com
Taylor & Francis Verlag GmbH, Kaufingerstraße 24, 80331 München, Germany

www.ingramcontent.com/pod-product-compliance
Ingram Content Group UK Ltd.
Pitfield, Milton Keynes, MK11 3LW, UK
UKHW020206200926
475702UK00004B/637
* 9 7 8 1 0 4 1 0 8 6 6 9 7 *